The Student's
Anatomy
of Exercise Manual

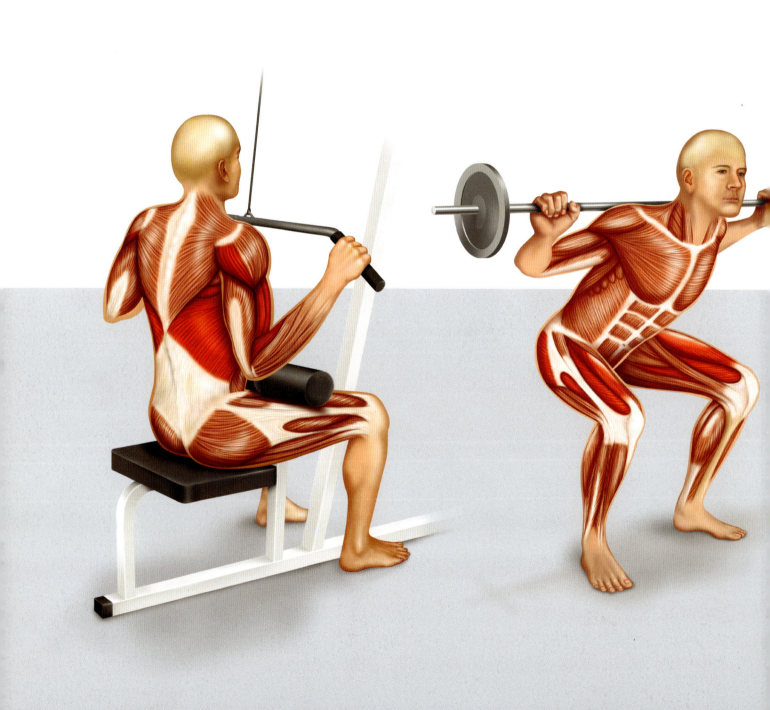

The Student's Anatomy of Exercise Manual

Professor Ken Ashwell: BMedSc, MBBS, PhD

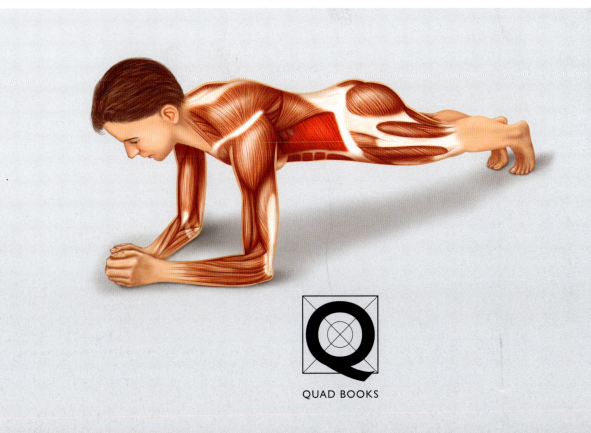

QUAD BOOKS

4

Quarto is the authority on a wide range of topics.

Quarto educates, entertains and enriches the lives of our readers—enthusiasts and lovers of hands-on living.

www.QuartoKnows.com

First published in 2014 by Global Book Publishing Pty Ltd
Global Book Publishing Pty Ltd
10.2 Level 10, 8 West Street,
North Sydney, 2060, Australia

ISBN 978-0-85762-469-7

Printed in China

Conceived, designed and produced by Global Book Publishing Pty Ltd

*It is recommended that anyone who is considering participating in an
exercise program should consult a physician before starting and that
no one should attempt a new exercise without the supervision of a
certified professional. While every care has been taken in presenting
this material, the anatomical and medical information is not intended
to replace professional medical advice; it should not be used as a
guide for self-treatment or self-diagnosis. Neither the authors nor the
publisher may be held responsible for any type of damage or harm
caused by the use or misuse of information in this book.*

Chief Consultant	Ken Ashwell B.Med.Sc., M.B.B.S., Ph.D.
Authors	Ken Ashwell B.Med.Sc., M.B.B.S., Ph.D.
	Michael Baker B.App.Sc., M.App.Sc., Ph.D., A.E.P.
	Tim Foulcher B.App.Sc., M.Phty.
	Michael Newton B.App.Sc., M.Sc., Ph.D., A.E.P.
Illustrator (Exercises)	Kristen W. Marzejon M.A.M.S., C.M.I.
Other Illustrations	David Carroll, Peter Child, Deborah Clarke, Geoff Cook, Marcus Cremonese, Beth Croce, Hans De Haas, Wendy de Paauw, Levant Efe, Mike Golding, Mike Gorman, Jeff Lang, Alex Lavroff, Ulrich Lehmann, Ruth Lindsay, Richard McKenna, Annabel Milne, Tony Pyrzakowski, Oliver Rennert, Caroline Rodrigues, Otto Schmidinger, Bob Seal, Vicky Short, Graeme Tavendale, Thomson Digital, Jonathan Tidball, Paul Tresnan, Valentin Varetsa, Glen Vause, Spike Wademan, Trevor Weekes, Paul Williams, David Wood

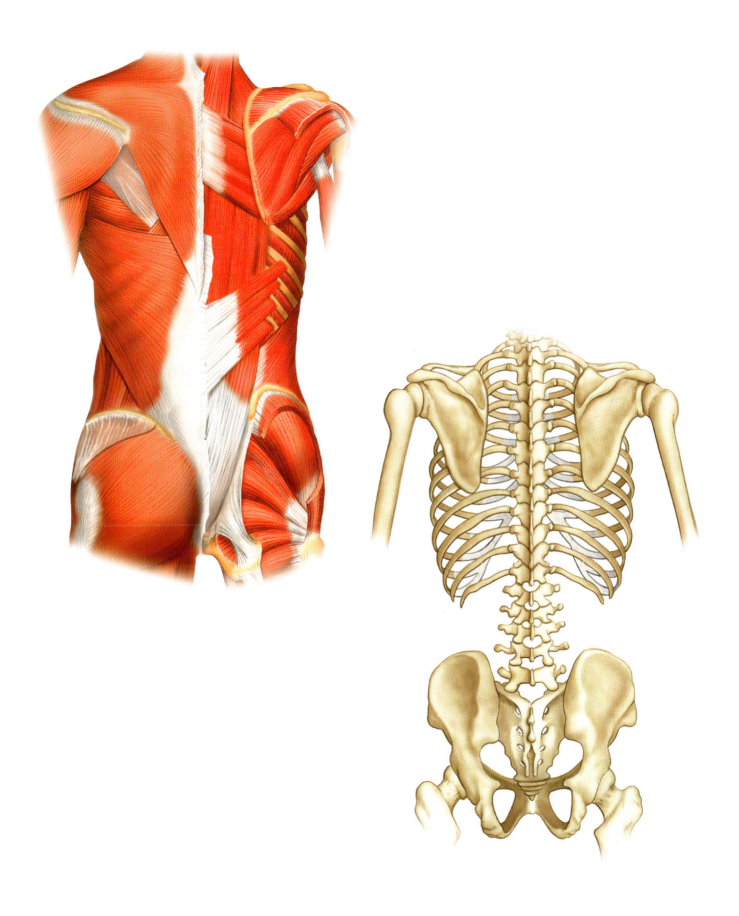

Contents

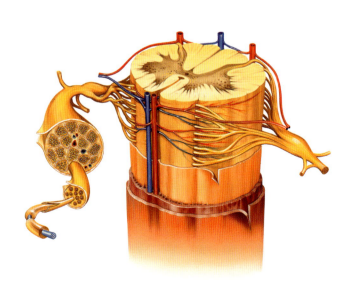

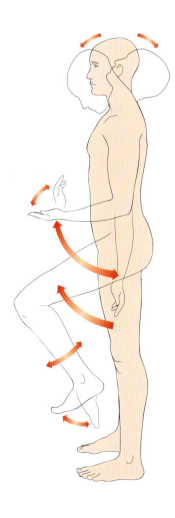

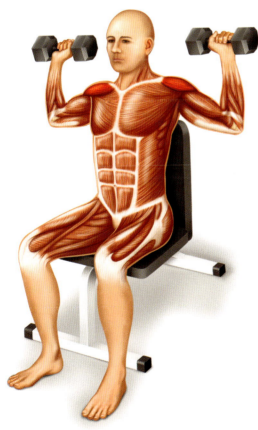

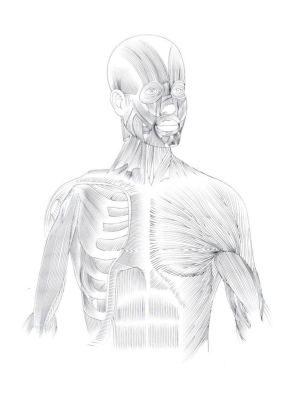

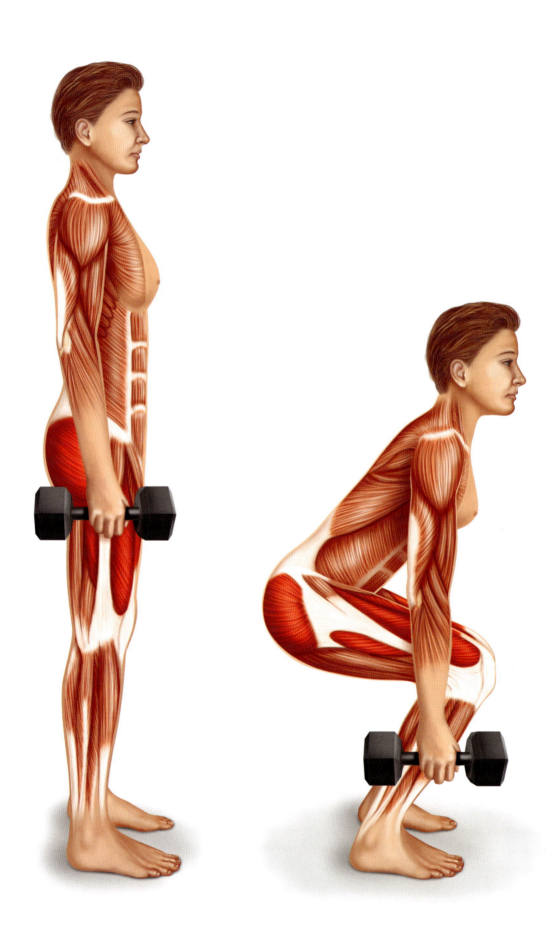

Foreword

There is a revolution in thinking about exercise. Exercise is no longer the sole domain of the sportsperson or body builder. Health professionals now know that physical activity is central to maintaining health and coordination for everyone from childhood to old age. Exercise is not a spectator sport and just buying a gym membership is not enough to keep your body in optimal shape. All adults should engage in an age-appropriate regime of exercise that takes their medical history and body type into account.

For the student of anatomy or sports physiology – who will train the athletes of tomorrow, as well as keeping all of the rest of us in shape – and the sportsperson, body builder, or just anyone wanting to improve their physical fitness, this book brings together sound practical advice on how to perform the most important exercises, with the anatomical detail to explain what each exercise is doing for the body. You will learn how specific exercises target particular muscle groups to ensure that you or your clients get the best result, whether for sporting needs or general fitness. The book includes an overview section at the front to explain the anatomy and function of key muscles, as well as a colouring workbook at the back to reinforce what you have learned.

Always remember to follow the advice and warnings for each exercise. Anyone beginning an exercise programme should first check with their medical practitioner, particularly if they are over 40 years old or have a history of heart disease, or high blood pressure. An exercise professional at your local gym will advise on the combination of exercises that are best for you. Some exercises use heavy weights and test the strength and flexibility of muscles that we do not often use in our daily lives. 'Do it right' means preparing properly for each exercise, using only the correct equipment, seeking help from an exercise professional or spotter where necessary, and considering the safety of those around you when performing the exercise. Safety for yourself and others should always be the highest priority.

Professor Ken Ashwell
Department of Anatomy,
School of Medical Sciences, Faculty of Medicine,
University of New South Wales, Sydney, Australia

How This Book Works

This book is organised into three primary sections: a full-colour anatomy overview; a full-colour illustrated exercise guide, comprising the main part of the book; and a colouring workbook in which to test your anatomical knowledge.

The anatomy overview section provides detailed, anatomically correct illustrations with clear, informative labels for the various body systems and regions. Visualising the parts of the body and their links to each other will improve your understanding of how the body works during exercise.

Each of the five chapters in the exercise guide focuses on specific muscle areas – arms and shoulders, chest, back, trunk, and legs and buttocks. Every exercise is depicted with two anatomically correct poses. Labels identify all the important muscles – including identifying active and stabiliser muscles – so you can visualise and understand exactly which muscles are activated during the exercises. This will not only increase your knowledge of anatomy; it will help to improve the effectiveness of workouts and rehabilitation programmemes.

The colouring workbook chapter is a study aid that aims to facilitate your understanding of important body systems – the muscular, skeletal, and nervous systems. Colour in each illustration to help memorise the location of muscles, bones, and nerves within these systems. Fill in the blank labels to test your knowledge of the names of body parts – the answers are given at the bottom of each page.

ANATOMY OVERVIEW PAGES

This section contains full-colour, double-page overview spreads that give a rundown on the important parts of a particular body system.

Section name

Subject title

20 · anatomy overview

muscular system · 21

Muscles of the Upper and Lower Limb

Superficial Muscles of the Upper Limb—Posterior View

Superficial Muscles of the Upper Limb—Anterior View

Superficial Muscles of the Lower Limb—Anterior View

Superficial Muscles of the Lower Limb—Posterior View

Labels

The labels on each illustration name the important parts of the body system, organ, or microstructure.

Illustration headings

Illustration headings give the name of the body part. The orientation is included where necessary.

EXERCISE PAGES

Each chapter in the exercise guide focuses on specific muscle areas. The exercises depict two anatomically correct poses that identify the active and stabiliser muscles.

Chapter name

Exercise title

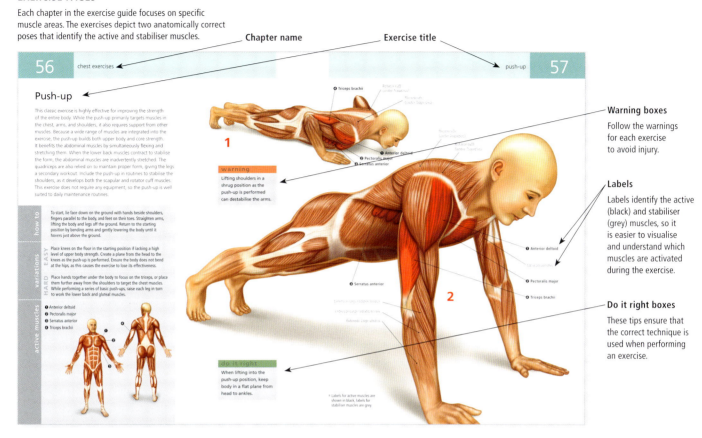

56   chest exercises

push-up   57

Push-up

This classic exercise is highly effective for improving the strength of the entire body. While the push-up primarily targets muscles in the chest, arms, and shoulders, it also requires support from other muscles. Because a wide range of muscles are integrated into the exercise, the push-up builds both upper body and core strength. It benefits the abdominal muscles by simultaneously flexing and stretching them. When the lower back muscles contract to stabilise the form, the abdominal muscles are inadvertently stretched. The quadriceps are also relied on to maintain proper form, giving the legs a secondary workout. Include the push-up in routines to stabilise the shoulders, as it develops both the scapular and rotator cuff muscles. This exercise does not require any equipment, so the push-up is well suited to daily maintenance routines.

how to
To start, lie face down on the ground with hands beside shoulders, fingers parallel to the body, and feet on their toes. Straighten arms, lifting the body and legs off the ground. Return to the starting position by bending arms and gently lowering the body until it hovers just above the ground.

variations
EASY
Place knees on the floor in the starting position if lacking a high level of upper body strength. Create a plane from the head to the knees as the push-up is performed. Ensure the body does not bend at the hips, as this causes the exercise to lose its effectiveness.

HARD
Place hands together under the body to focus on the triceps, or place them further away from the shoulders to target the chest muscles. While performing a series of basic push-ups, raise each leg in turn to work the lower back and gluteal muscles.

active muscles
❶ Anterior deltoid
❷ Pectoralis major
❸ Serratus anterior
❹ Triceps brachii

1

warning
Lifting shoulders in a shrug position as the push-up is performed can destabilise the arms.

❹ Triceps brachii

❸ Anterior deltoid
❷ Pectoralis major
❶ Serratus anterior

2

❷ Serratus anterior

❶ Anterior deltoid

❷ Pectoralis major

❹ Triceps brachii

do it right
When lifting into the push-up position, keep body in a flat plane from head to ankles.

* Labels for active muscles are shown in black, labels for stabiliser muscles are grey.

Warning boxes
Follow the warnings for each exercise to avoid injury.

Labels
Labels identify the active (black) and stabiliser (grey) muscles, so it is easier to visualise and understand which muscles are activated during the exercise.

Do it right boxes
These tips ensure that the correct technique is used when performing an exercise.

COLOURING WORKBOOK PAGES

This final section contains black-and-white drawings of parts of the muscular, skeletal, and nervous systems. Colour in the body parts as a memory aid.

Section name

Subject title

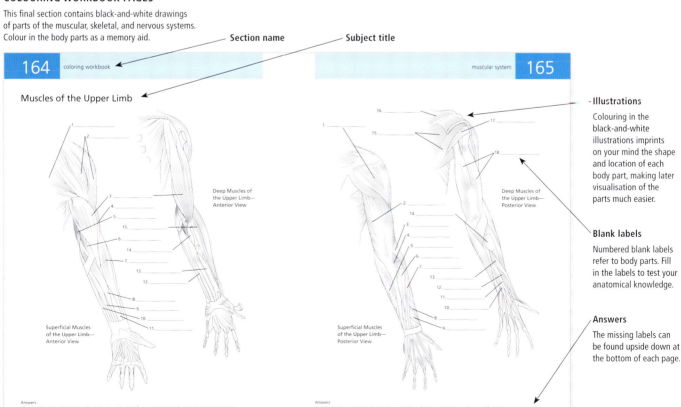

164   coloring workbook

muscular system   165

Muscles of the Upper Limb

Deep Muscles of the Upper Limb— Anterior View

Superficial Muscles of the Upper Limb— Anterior View

Deep Muscles of the Upper Limb— Posterior View

Superficial Muscles of the Upper Limb— Posterior View

Illustrations
Colouring in the black-and-white illustrations imprints on your mind the shape and location of each body part, making later visualisation of the parts much easier.

Blank labels
Numbered blank labels refer to body parts. Fill in the labels to test your anatomical knowledge.

Answers
The missing labels can be found upside down at the bottom of each page.

Answers

Answers

Anatomy Overview

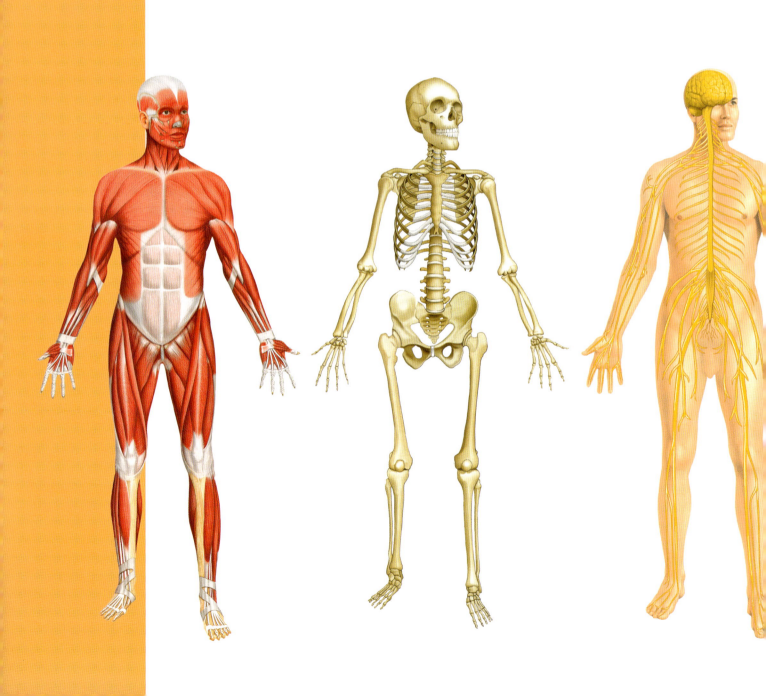

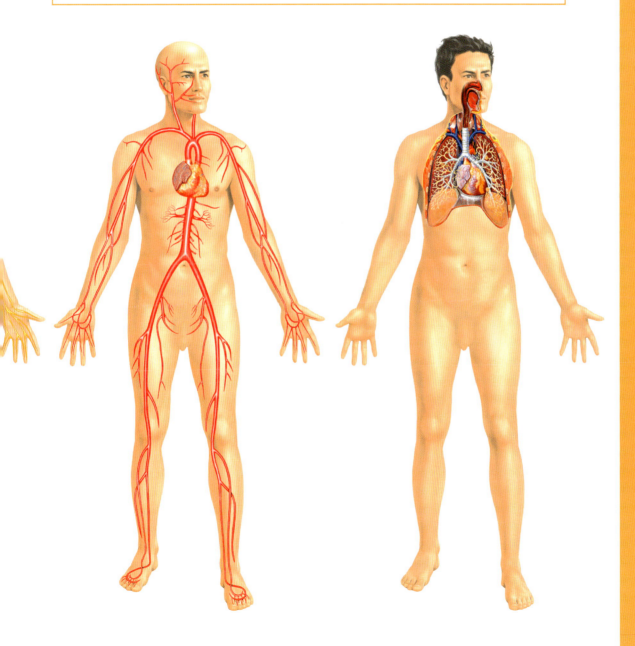

Body Regions

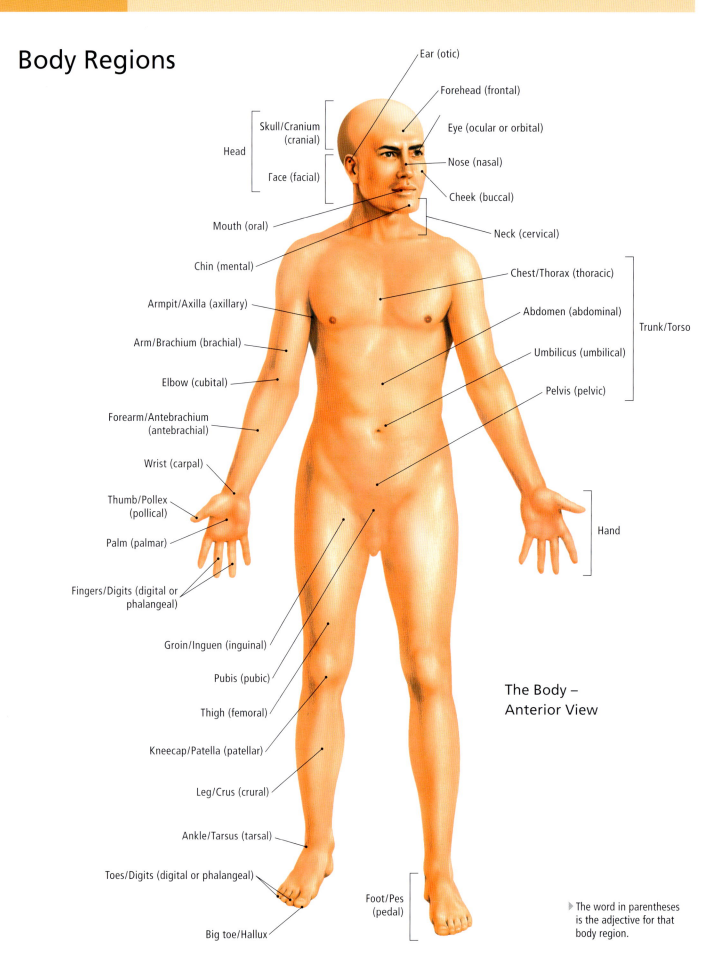

Ear (otic)

Forehead (frontal)

Eye (ocular or orbital)

Nose (nasal)

Cheek (buccal)

Neck (cervical)

Head

Skull/Cranium (cranial)

Face (facial)

Mouth (oral)

Chin (mental)

Armpit/Axilla (axillary)

Arm/Brachium (brachial)

Elbow (cubital)

Forearm/Antebrachium (antebrachial)

Wrist (carpal)

Thumb/Pollex (pollical)

Palm (palmar)

Fingers/Digits (digital or phalangeal)

Groin/Inguen (inguinal)

Pubis (pubic)

Thigh (femoral)

Kneecap/Patella (patellar)

Leg/Crus (crural)

Ankle/Tarsus (tarsal)

Toes/Digits (digital or phalangeal)

Big toe/Hallux

Chest/Thorax (thoracic)

Abdomen (abdominal)

Umbilicus (umbilical)

Pelvis (pelvic)

Trunk/Torso

Hand

The Body –
Anterior View

Foot/Pes (pedal)

▶ The word in parentheses is the adjective for that body region.

The Body –
Posterior View

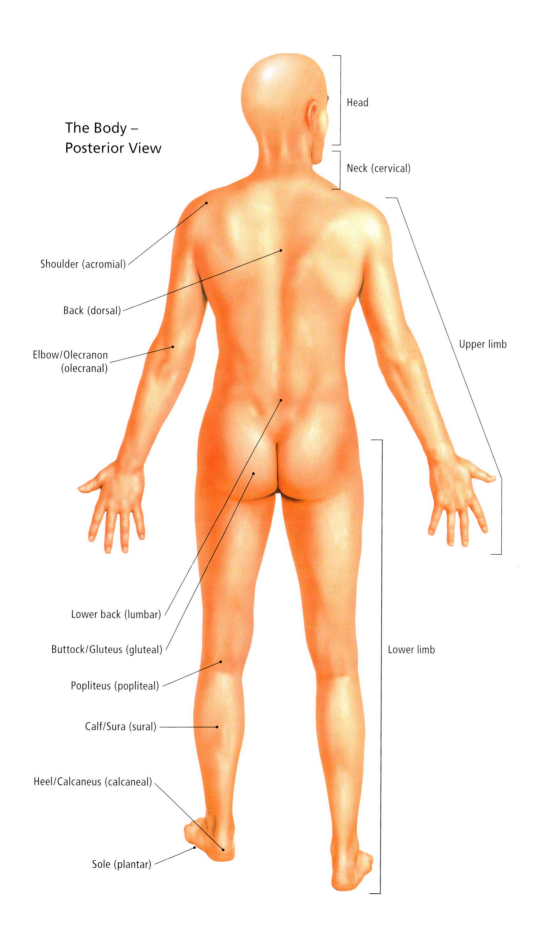

Head

Neck (cervical)

Shoulder (acromial)

Back (dorsal)

Elbow/Olecranon
(olecranal)

Upper limb

Lower back (lumbar)

Buttock/Gluteus (gluteal)

Popliteus (popliteal)

Lower limb

Calf/Sura (sural)

Heel/Calcaneus (calcaneal)

Sole (plantar)

Muscles of the Body

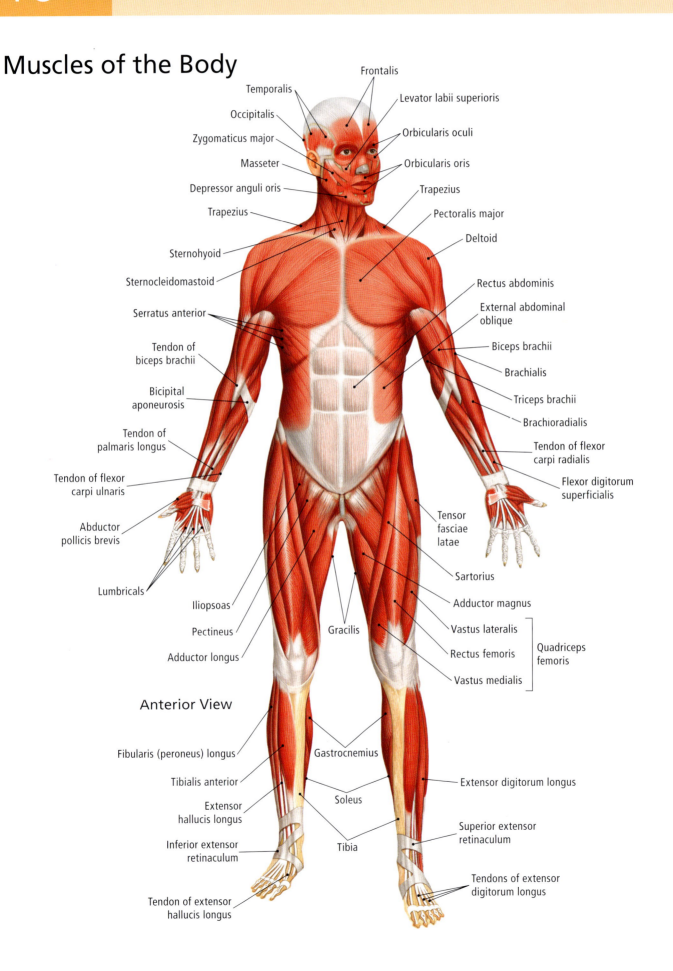

Frontalis

Temporalis

Levator labii superioris

Occipitalis

Orbicularis oculi

Zygomaticus major

Orbicularis oris

Masseter

Depressor anguli oris

Trapezius

Trapezius

Pectoralis major

Sternohyoid

Deltoid

Sternocleidomastoid

Rectus abdominis

Serratus anterior

External abdominal oblique

Tendon of biceps brachii

Biceps brachii

Brachialis

Bicipital aponeurosis

Triceps brachii

Brachioradialis

Tendon of palmaris longus

Tendon of flexor carpi radialis

Tendon of flexor carpi ulnaris

Flexor digitorum superficialis

Abductor pollicis brevis

Tensor fasciae latae

Lumbricals

Sartorius

Iliopsoas

Adductor magnus

Pectineus

Gracilis

Vastus lateralis

Adductor longus

Rectus femoris

Quadriceps femoris

Vastus medialis

Anterior View

Fibularis (peroneus) longus

Gastrocnemius

Tibialis anterior

Extensor digitorum longus

Extensor hallucis longus

Soleus

Superior extensor retinaculum

Inferior extensor retinaculum

Tibia

Tendon of extensor hallucis longus

Tendons of extensor digitorum longus

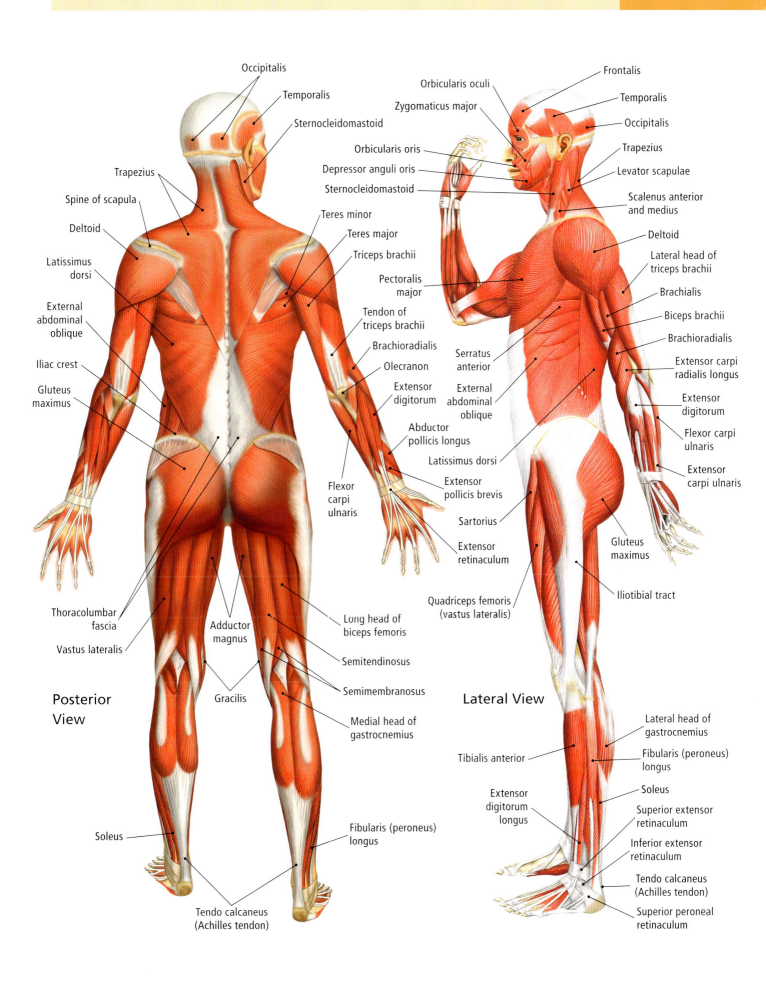

Occipitalis

Temporalis

Sternocleidomastoid

Orbicularis oculi

Zygomaticus major

Frontalis

Temporalis

Occipitalis

Trapezius

Levator scapulae

Orbicularis oris

Depressor anguli oris

Sternocleidomastoid

Scalenus anterior
and medius

Trapezius

Spine of scapula

Deltoid

Latissimus
dorsi

External
abdominal
oblique

Iliac crest

Gluteus
maximus

Teres minor

Teres major

Triceps brachii

Pectoralis
major

Tendon of
triceps brachii

Brachioradialis

Olecranon

Extensor
digitorum

Abductor
pollicis longus

Serratus
anterior

External
abdominal
oblique

Latissimus dorsi

Extensor
pollicis brevis

Sartorius

Extensor
retinaculum

Deltoid

Lateral head of
triceps brachii

Brachialis

Biceps brachii

Brachioradialis

Extensor carpi
radialis longus

Extensor
digitorum

Flexor carpi
ulnaris

Extensor
carpi ulnaris

Gluteus
maximus

Iliotibial tract

Flexor
carpi
ulnaris

Quadriceps femoris
(vastus lateralis)

Thoracolumbar
fascia

Vastus lateralis

Adductor
magnus

Gracilis

Long head of
biceps femoris

Semitendinosus

Semimembranosus

Medial head of
gastrocnemius

Posterior
View

Lateral View

Lateral head of
gastrocnemius

Fibularis (peroneus)
longus

Soleus

Superior extensor
retinaculum

Inferior extensor
retinaculum

Tendo calcaneus
(Achilles tendon)

Superior peroneal
retinaculum

Tibialis anterior

Extensor
digitorum
longus

Soleus

Fibularis (peroneus)
longus

Tendo calcaneus
(Achilles tendon)

Muscles of the Abdomen and Back

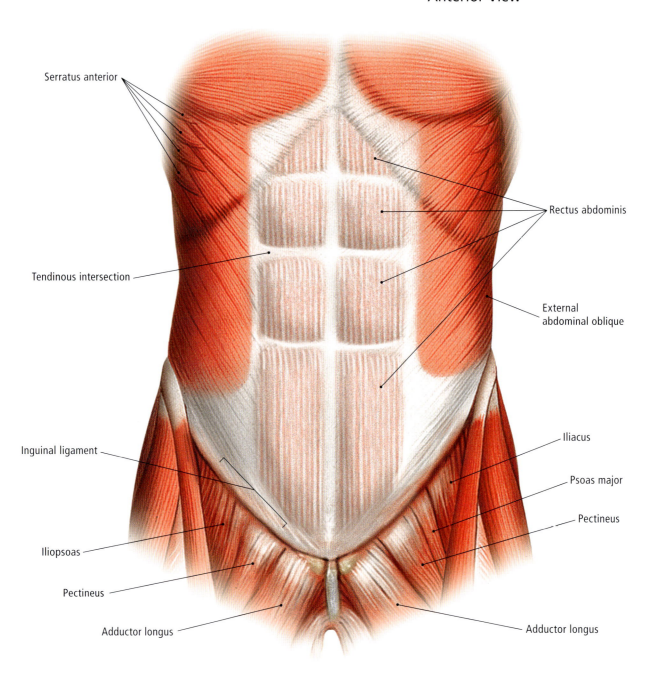

Muscles of the Abdomen –
Anterior View

Serratus anterior

Tendinous intersection

Inguinal ligament

Iliopsoas

Pectineus

Adductor longus

Rectus abdominis

External
abdominal oblique

Iliacus

Psoas major

Pectineus

Adductor longus

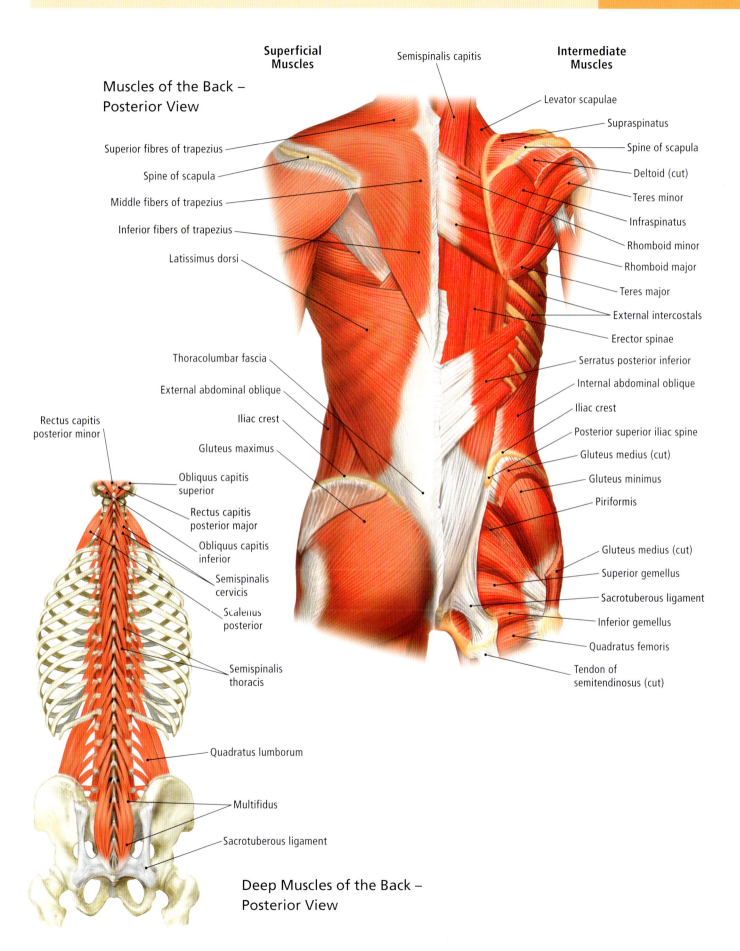

Muscles of the Back – Posterior View

Superficial Muscles

Semispinalis capitis

Intermediate Muscles

Superior fibres of trapezius

Spine of scapula

Middle fibers of trapezius

Inferior fibers of trapezius

Latissimus dorsi

Levator scapulae

Supraspinatus

Spine of scapula

Deltoid (cut)

Teres minor

Infraspinatus

Rhomboid minor

Rhomboid major

Teres major

External intercostals

Erector spinae

Thoracolumbar fascia

External abdominal oblique

Iliac crest

Gluteus maximus

Serratus posterior inferior

Internal abdominal oblique

Iliac crest

Posterior superior iliac spine

Gluteus medius (cut)

Gluteus minimus

Piriformis

Gluteus medius (cut)

Superior gemellus

Sacrotuberous ligament

Inferior gemellus

Quadratus femoris

Tendon of semitendinosus (cut)

Rectus capitis posterior minor

Obliquus capitis superior

Rectus capitis posterior major

Obliquus capitis inferior

Semispinalis cervicis

Scalenus posterior

Semispinalis thoracis

Quadratus lumborum

Multifidus

Sacrotuberous ligament

Deep Muscles of the Back – Posterior View

Muscles of the Upper and Lower Limb

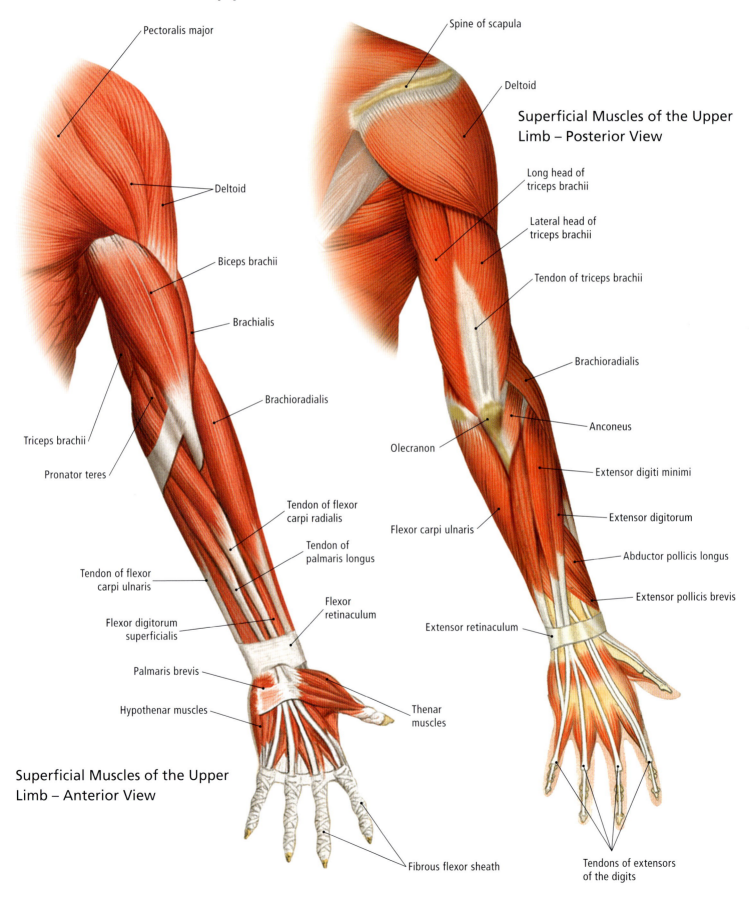

Superficial Muscles of the Upper Limb – Posterior View

Superficial Muscles of the Upper Limb – Anterior View

Labels (anterior view):
- Pectoralis major
- Deltoid
- Biceps brachii
- Brachialis
- Brachioradialis
- Triceps brachii
- Pronator teres
- Tendon of flexor carpi radialis
- Tendon of palmaris longus
- Tendon of flexor carpi ulnaris
- Flexor retinaculum
- Flexor digitorum superficialis
- Palmaris brevis
- Hypothenar muscles
- Thenar muscles
- Fibrous flexor sheath

Labels (posterior view):
- Spine of scapula
- Deltoid
- Long head of triceps brachii
- Lateral head of triceps brachii
- Tendon of triceps brachii
- Brachioradialis
- Anconeus
- Olecranon
- Extensor digiti minimi
- Flexor carpi ulnaris
- Extensor digitorum
- Abductor pollicis longus
- Extensor pollicis brevis
- Extensor retinaculum
- Tendons of extensors of the digits

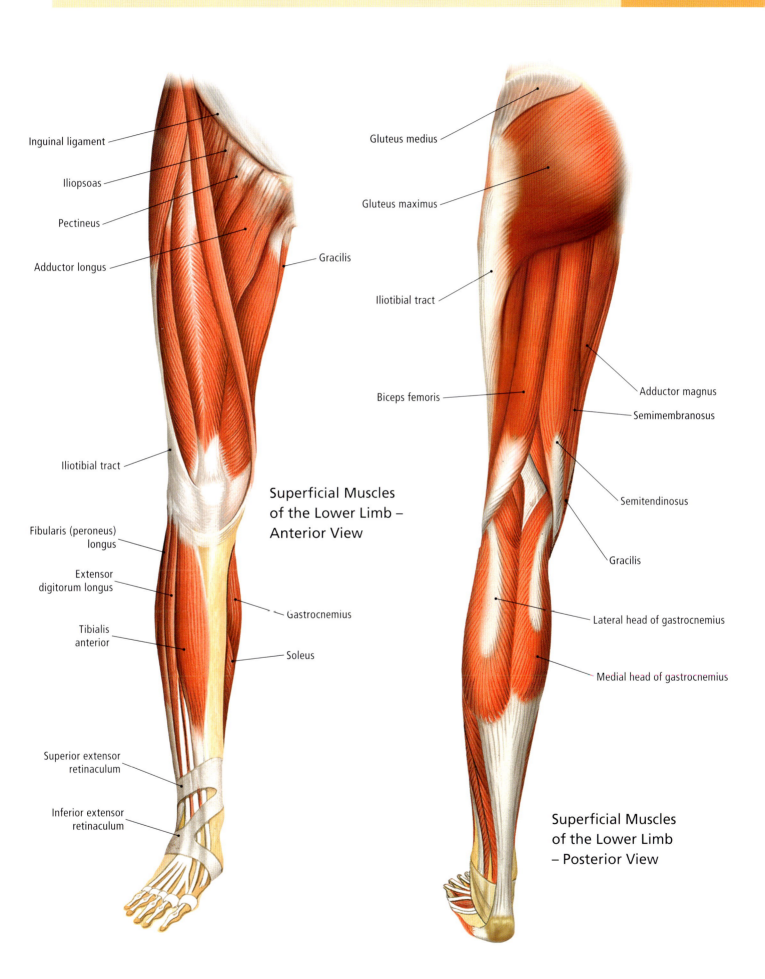

Inguinal ligament

Iliopsoas

Pectineus

Adductor longus

Gracilis

Iliotibial tract

Fibularis (peroneus) longus

Extensor digitorum longus

Tibialis anterior

Gastrocnemius

Soleus

Superior extensor retinaculum

Inferior extensor retinaculum

Superficial Muscles of the Lower Limb – Anterior View

Gluteus medius

Gluteus maximus

Iliotibial tract

Biceps femoris

Adductor magnus

Semimembranosus

Semitendinosus

Gracilis

Lateral head of gastrocnemius

Medial head of gastrocnemius

Superficial Muscles of the Lower Limb – Posterior View

Bones of the Body

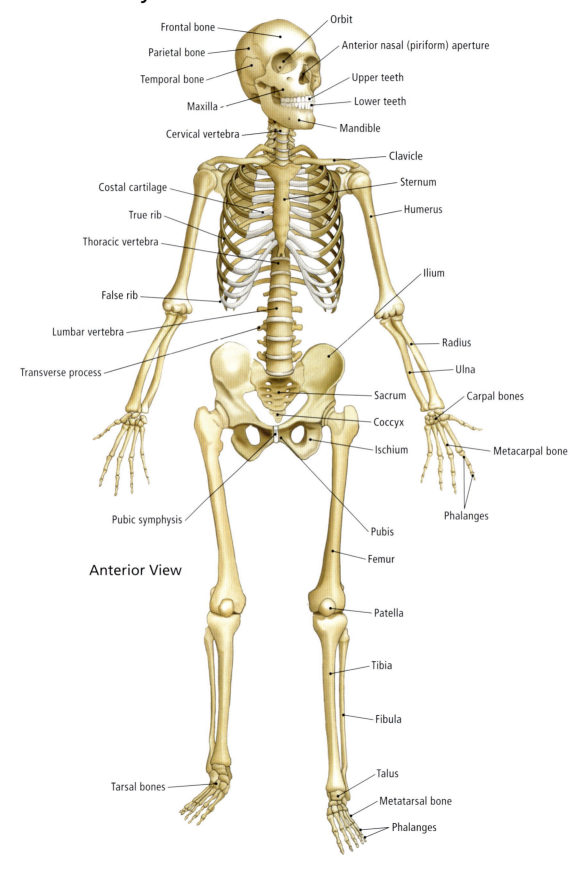

Frontal bone
Orbit
Parietal bone
Anterior nasal (piriform) aperture
Temporal bone
Upper teeth
Lower teeth
Maxilla
Mandible
Cervical vertebra
Clavicle
Costal cartilage
Sternum
True rib
Humerus
Thoracic vertebra
Ilium
False rib
Lumbar vertebra
Radius
Ulna
Transverse process
Sacrum
Carpal bones
Coccyx
Ischium
Metacarpal bone
Phalanges
Pubic symphysis
Pubis
Anterior View
Femur
Patella
Tibia
Fibula
Talus
Tarsal bones
Metatarsal bone
Phalanges

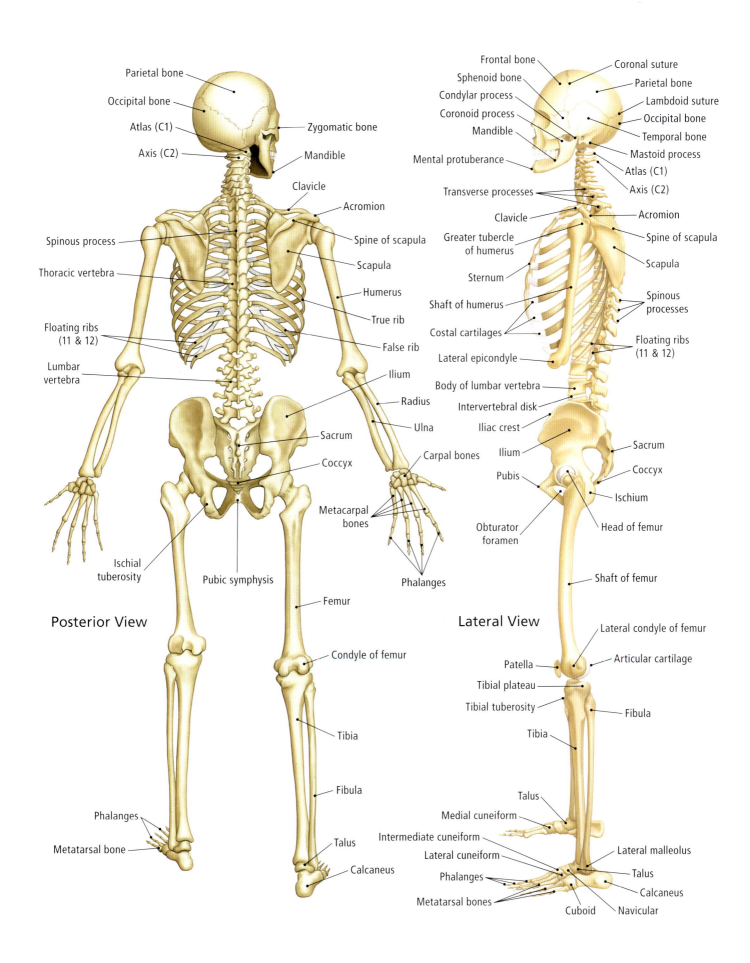

Parietal bone

Occipital bone

Atlas (C1)

Axis (C2)

Zygomatic bone

Mandible

Clavicle

Acromion

Spine of scapula

Spinous process

Scapula

Thoracic vertebra

Humerus

True rib

Floating ribs
(11 & 12)

False rib

Lumbar
vertebra

Ilium

Radius

Ulna

Sacrum

Coccyx

Carpal bones

Metacarpal
bones

Ischial
tuberosity

Pubic symphysis

Phalanges

Femur

Posterior View

Condyle of femur

Tibia

Fibula

Phalanges

Metatarsal bone

Talus

Calcaneus

Frontal bone

Sphenoid bone

Condylar process

Coronoid process

Mandible

Mental protuberance

Coronal suture

Parietal bone

Lambdoid suture

Occipital bone

Temporal bone

Mastoid process

Atlas (C1)

Axis (C2)

Transverse processes

Clavicle

Greater tubercle
of humerus

Sternum

Shaft of humerus

Costal cartilages

Lateral epicondyle

Body of lumbar vertebra

Intervertebral disk

Iliac crest

Ilium

Pubis

Obturator
foramen

Acromion

Spine of scapula

Scapula

Spinous
processes

Floating ribs
(11 & 12)

Sacrum

Coccyx

Ischium

Head of femur

Shaft of femur

Lateral View

Lateral condyle of femur

Patella

Tibial plateau

Tibial tuberosity

Articular cartilage

Fibula

Tibia

Talus

Medial cuneiform

Intermediate cuneiform

Lateral cuneiform

Phalanges

Metatarsal bones

Cuboid

Lateral malleolus

Talus

Calcaneus

Navicular

Vertebral Column

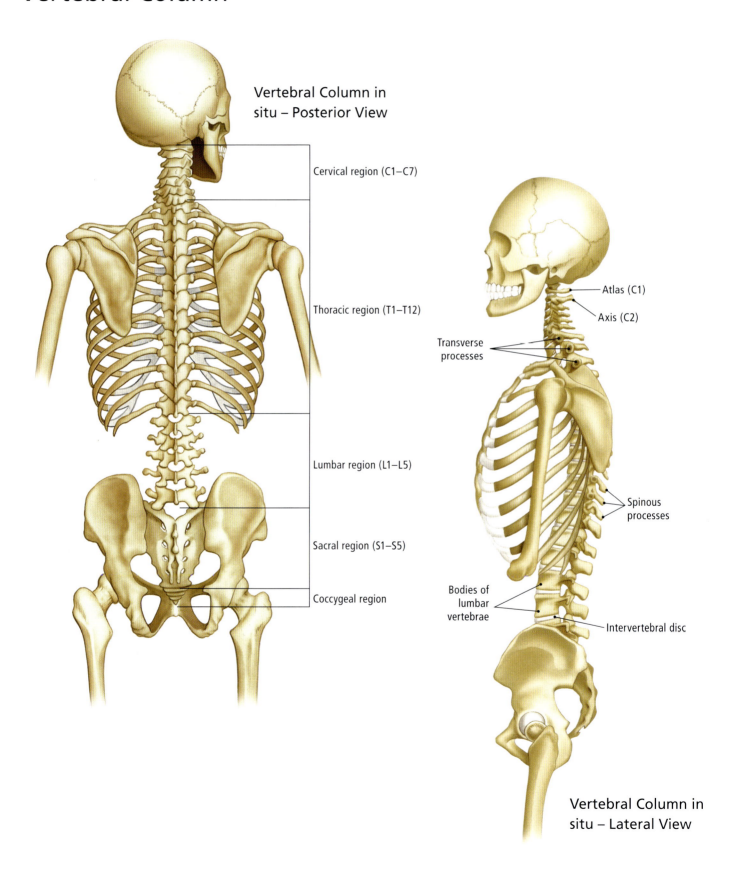

Vertebral Column in situ – Posterior View

Cervical region (C1–C7)

Thoracic region (T1–T12)

Lumbar region (L1–L5)

Sacral region (S1–S5)

Coccygeal region

Atlas (C1)

Axis (C2)

Transverse processes

Spinous processes

Bodies of lumbar vertebrae

Intervertebral disc

Vertebral Column in situ – Lateral View

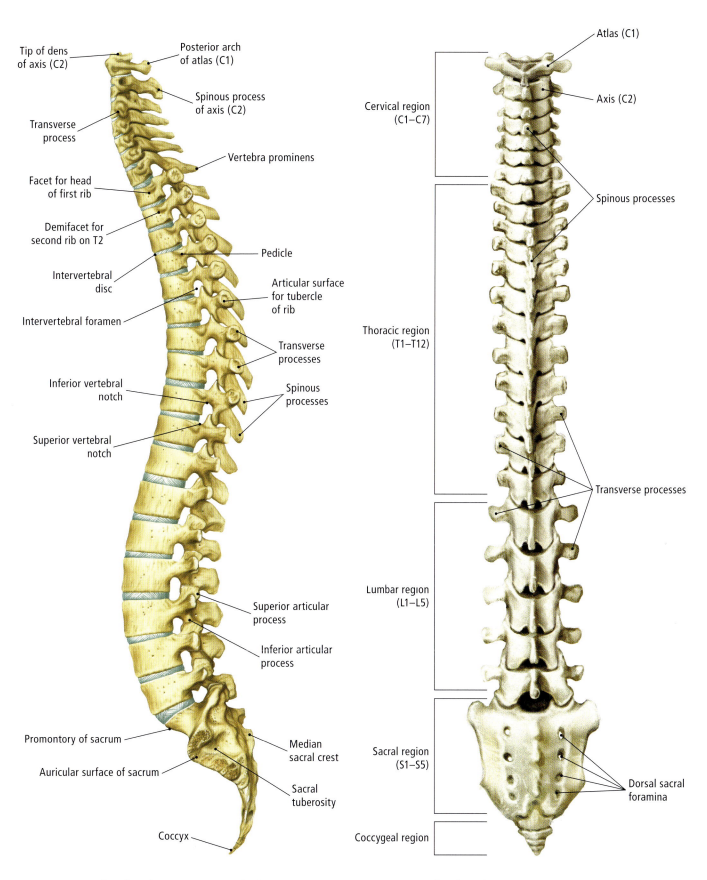

Tip of dens of axis (C2)

Posterior arch of atlas (C1)

Spinous process of axis (C2)

Transverse process

Vertebra prominens

Facet for head of first rib

Demifacet for second rib on T2

Pedicle

Intervertebral disc

Articular surface for tubercle of rib

Intervertebral foramen

Transverse processes

Inferior vertebral notch

Spinous processes

Superior vertebral notch

Superior articular process

Inferior articular process

Promontory of sacrum

Median sacral crest

Auricular surface of sacrum

Sacral tuberosity

Coccyx

Atlas (C1)

Axis (C2)

Cervical region (C1–C7)

Spinous processes

Thoracic region (T1–T12)

Transverse processes

Lumbar region (L1–L5)

Sacral region (S1–S5)

Dorsal sacral foramina

Coccygeal region

Vertebral Column – Lateral View

Vertebral Column – Posterior View

Bones of the Upper and Lower Limb

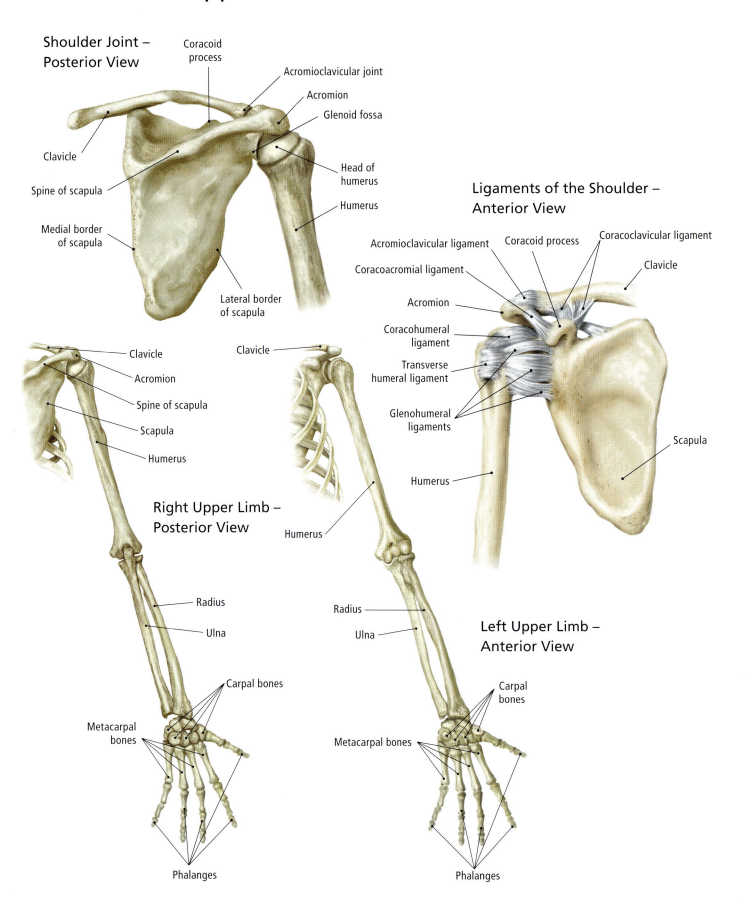

Shoulder Joint – Posterior View

- Coracoid process
- Acromioclavicular joint
- Acromion
- Glenoid fossa
- Clavicle
- Head of humerus
- Spine of scapula
- Humerus
- Medial border of scapula
- Lateral border of scapula

Ligaments of the Shoulder – Anterior View

- Acromioclavicular ligament
- Coracoid process
- Coracoclavicular ligament
- Coracoacromial ligament
- Clavicle
- Acromion
- Coracohumeral ligament
- Transverse humeral ligament
- Glenohumeral ligaments
- Humerus
- Scapula

Right Upper Limb – Posterior View

- Clavicle
- Acromion
- Spine of scapula
- Scapula
- Humerus
- Radius
- Ulna
- Carpal bones
- Metacarpal bones
- Phalanges

Left Upper Limb – Anterior View

- Clavicle
- Humerus
- Radius
- Ulna
- Carpal bones
- Metacarpal bones
- Phalanges

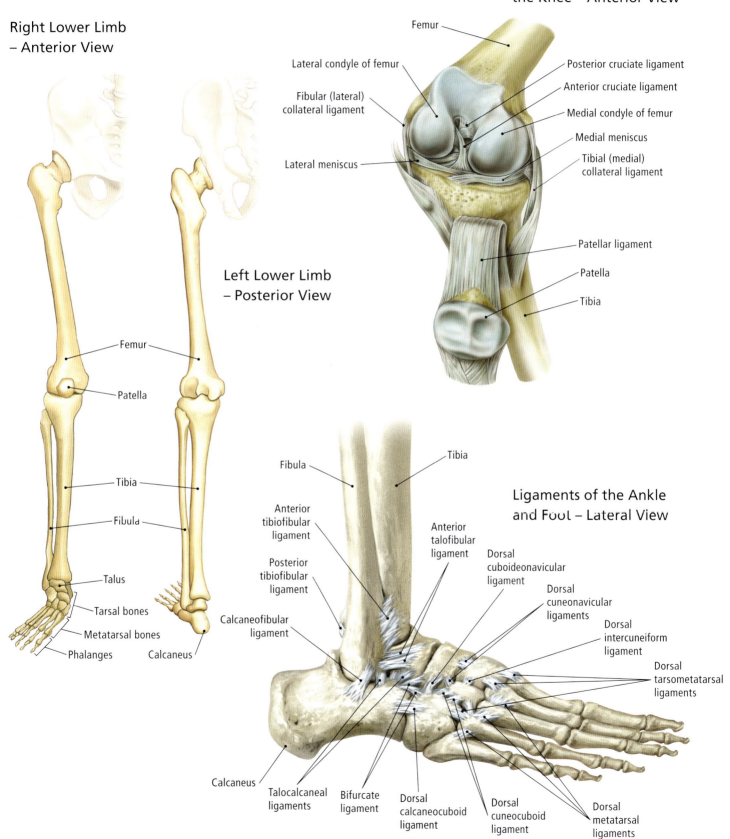

Bones and Ligaments of the Knee – Anterior View

Femur

Lateral condyle of femur

Fibular (lateral) collateral ligament

Lateral meniscus

Posterior cruciate ligament

Anterior cruciate ligament

Medial condyle of femur

Medial meniscus

Tibial (medial) collateral ligament

Patellar ligament

Patella

Tibia

Right Lower Limb – Anterior View

Left Lower Limb – Posterior View

Femur

Patella

Tibia

Fibula

Talus

Tarsal bones

Metatarsal bones

Phalanges

Calcaneus

Fibula

Tibia

Anterior tibiofibular ligament

Posterior tibiofibular ligament

Calcaneofibular ligament

Anterior talofibular ligament

Dorsal cuboideonavicular ligament

Dorsal cuneonavicular ligaments

Dorsal intercuneiform ligament

Dorsal tarsometatarsal ligaments

Ligaments of the Ankle and Foot – Lateral View

Calcaneus

Talocalcaneal ligaments

Bifurcate ligament

Dorsal calcaneocuboid ligament

Dorsal cuneocuboid ligament

Dorsal metatarsal ligaments

Nervous System

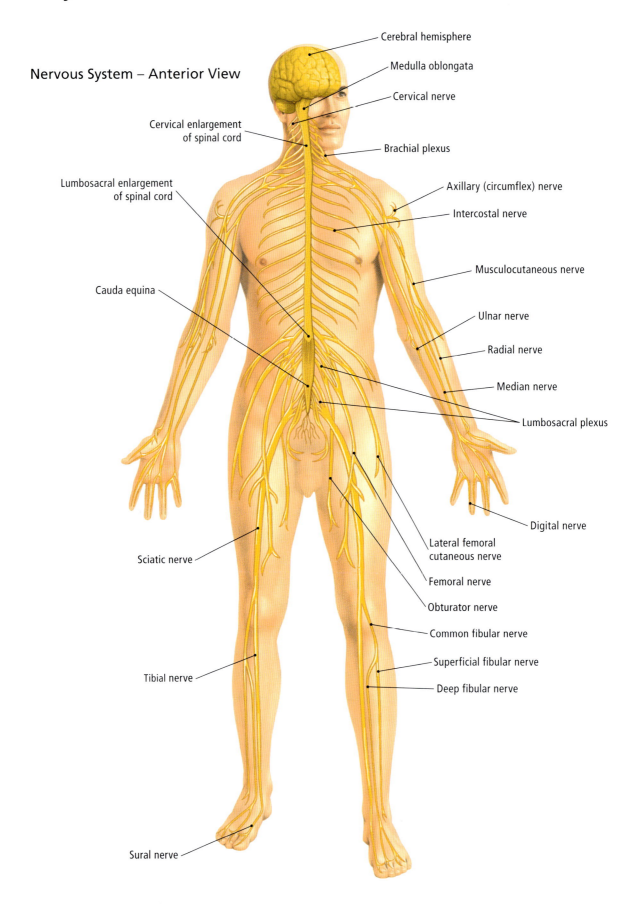

Nervous System – Anterior View

Cerebral hemisphere
Medulla oblongata
Cervical nerve
Cervical enlargement of spinal cord
Brachial plexus
Lumbosacral enlargement of spinal cord
Axillary (circumflex) nerve
Intercostal nerve
Musculocutaneous nerve
Cauda equina
Ulnar nerve
Radial nerve
Median nerve
Lumbosacral plexus
Digital nerve
Lateral femoral cutaneous nerve
Sciatic nerve
Femoral nerve
Obturator nerve
Common fibular nerve
Tibial nerve
Superficial fibular nerve
Deep fibular nerve
Sural nerve

Central Nervous System

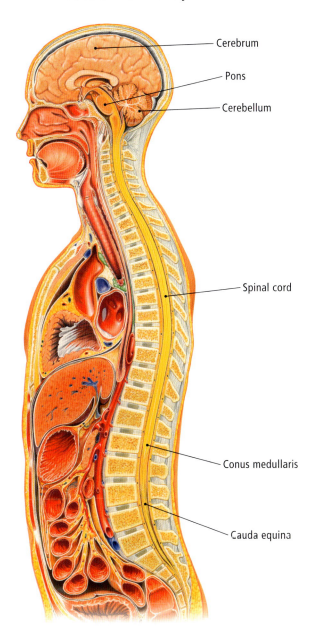

Cerebrum

Pons

Cerebellum

Spinal cord

Conus medullaris

Cauda equina

Autonomic Nervous System

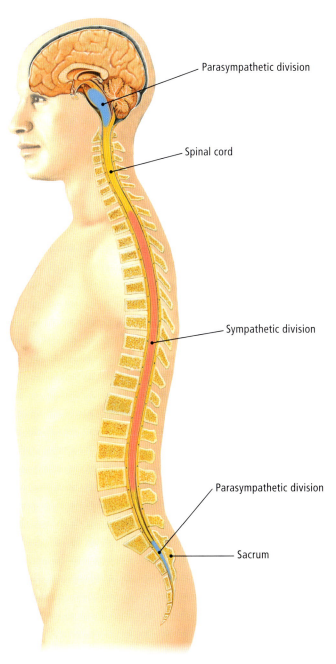

Parasympathetic division

Spinal cord

Sympathetic division

Parasympathetic division

Sacrum

Spinal Cord

Spinal Cord – Cross-sectional View

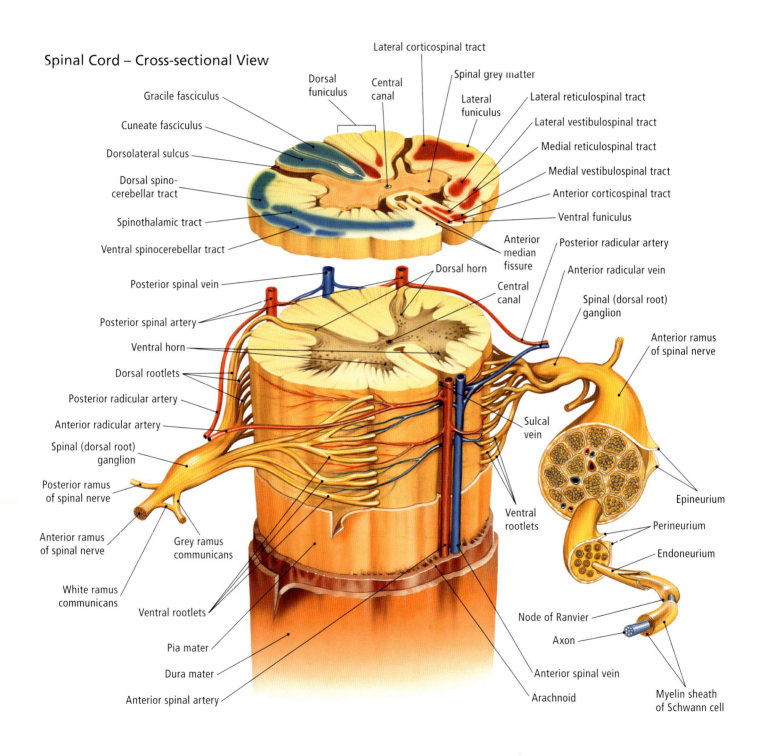

Lateral corticospinal tract

Dorsal funiculus

Central canal

Spinal grey matter

Lateral funiculus

Gracile fasciculus

Cuneate fasciculus

Dorsolateral sulcus

Dorsal spino-cerebellar tract

Spinothalamic tract

Ventral spinocerebellar tract

Lateral reticulospinal tract

Lateral vestibulospinal tract

Medial reticulospinal tract

Medial vestibulospinal tract

Anterior corticospinal tract

Ventral funiculus

Anterior median fissure

Posterior radicular artery

Dorsal horn

Central canal

Anterior radicular vein

Spinal (dorsal root) ganglion

Anterior ramus of spinal nerve

Posterior spinal vein

Posterior spinal artery

Ventral horn

Dorsal rootlets

Posterior radicular artery

Anterior radicular artery

Spinal (dorsal root) ganglion

Posterior ramus of spinal nerve

Anterior ramus of spinal nerve

White ramus communicans

Grey ramus communicans

Ventral rootlets

Pia mater

Dura mater

Anterior spinal artery

Sulcal vein

Ventral rootlets

Epineurium

Perineurium

Endoneurium

Node of Ranvier

Axon

Anterior spinal vein

Arachnoid

Myelin sheath of Schwann cell

Spinal Nerves

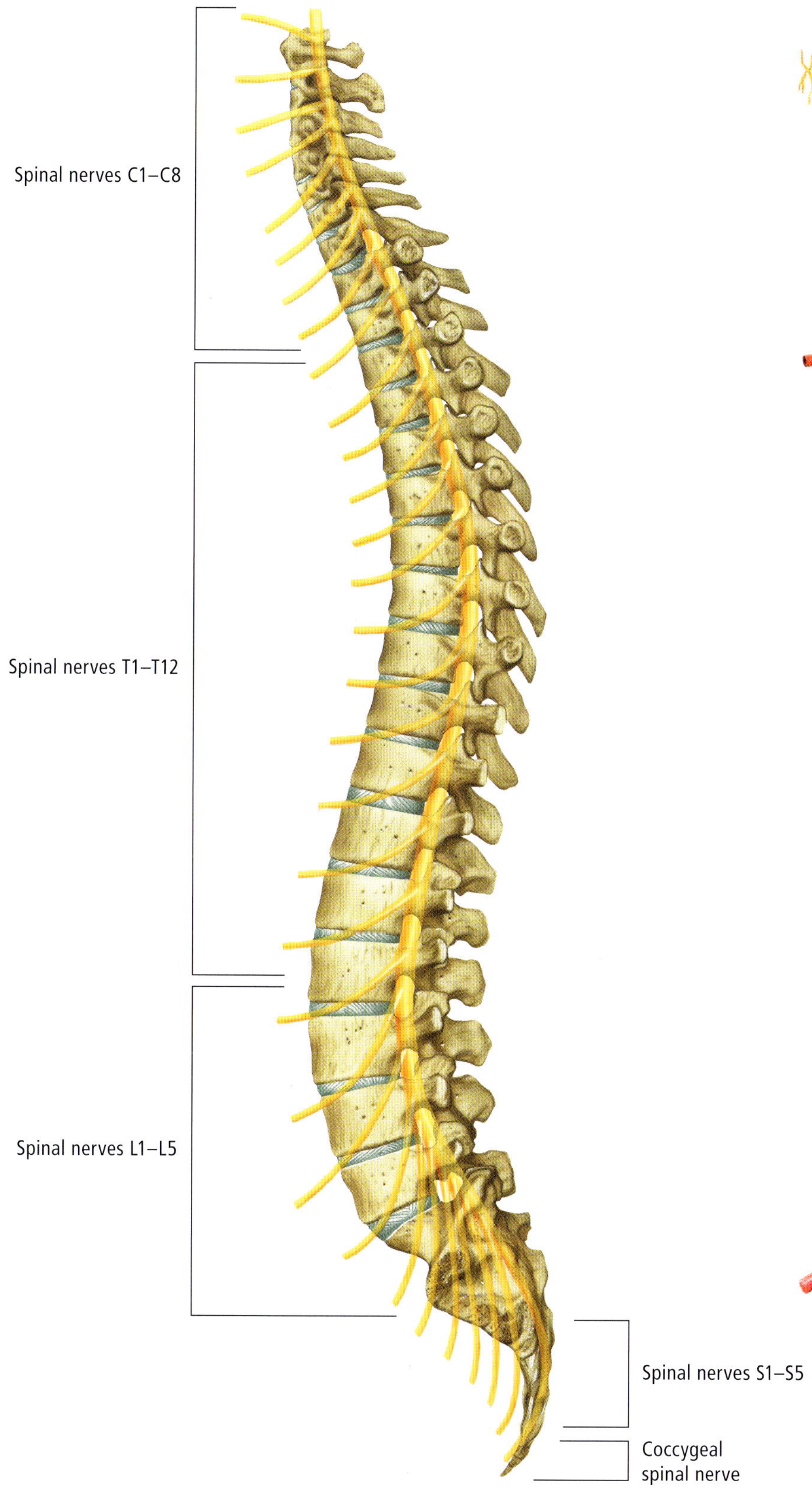

Spinal nerves C1–C8

Spinal nerves T1–T12

Spinal nerves L1–L5

Spinal nerves S1–S5

Coccygeal
spinal nerve

Spinal Cord – Anterior View

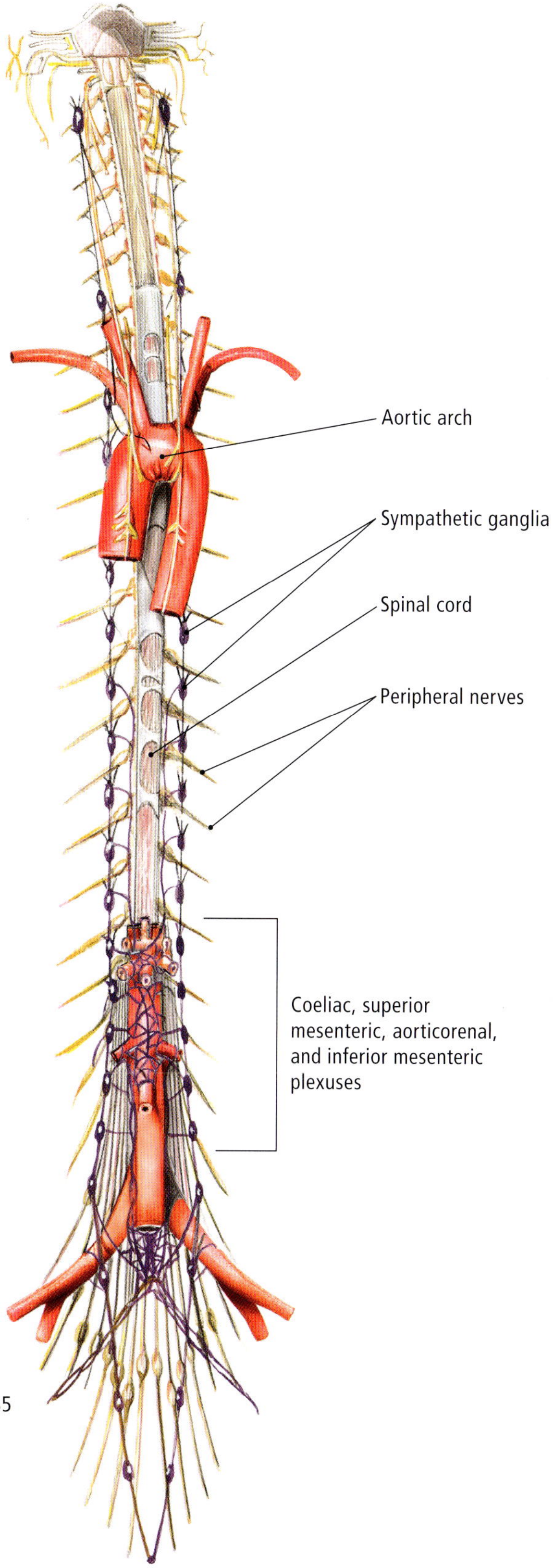

Aortic arch

Sympathetic ganglia

Spinal cord

Peripheral nerves

Coeliac, superior
mesenteric, aorticorenal,
and inferior mesenteric
plexuses

Circulatory System

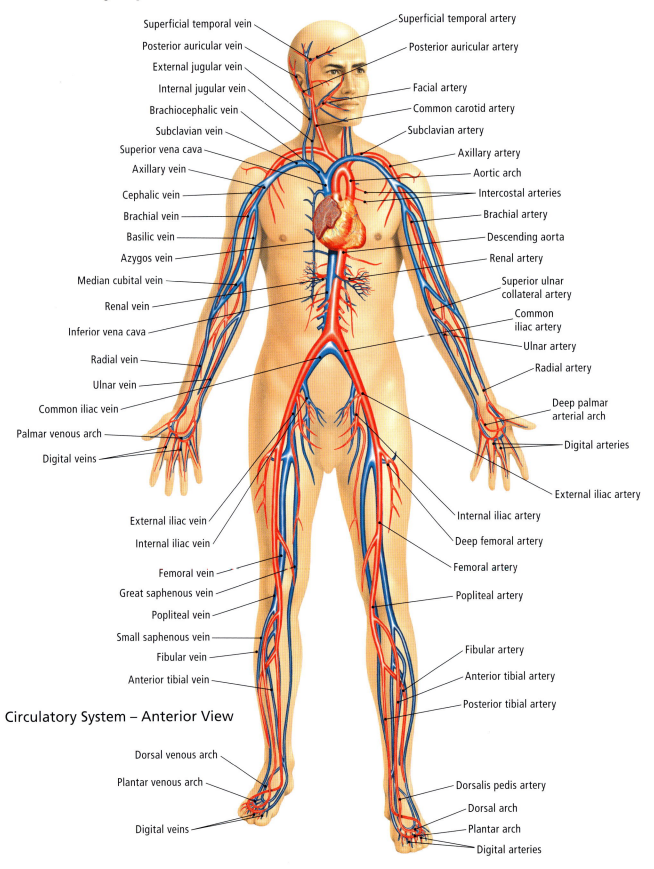

Superficial temporal vein

Posterior auricular vein

External jugular vein

Internal jugular vein

Brachiocephalic vein

Subclavian vein

Superior vena cava

Axillary vein

Cephalic vein

Brachial vein

Basilic vein

Azygos vein

Median cubital vein

Renal vein

Inferior vena cava

Radial vein

Ulnar vein

Common iliac vein

Palmar venous arch

Digital veins

Superficial temporal artery

Posterior auricular artery

Facial artery

Common carotid artery

Subclavian artery

Axillary artery

Aortic arch

Intercostal arteries

Brachial artery

Descending aorta

Renal artery

Superior ulnar collateral artery

Common iliac artery

Ulnar artery

Radial artery

Deep palmar arterial arch

Digital arteries

External iliac artery

Internal iliac artery

Deep femoral artery

Femoral artery

Popliteal artery

External iliac vein

Internal iliac vein

Femoral vein

Great saphenous vein

Popliteal vein

Small saphenous vein

Fibular vein

Anterior tibial vein

Fibular artery

Anterior tibial artery

Posterior tibial artery

Circulatory System – Anterior View

Dorsal venous arch

Plantar venous arch

Digital veins

Dorsalis pedis artery

Dorsal arch

Plantar arch

Digital arteries

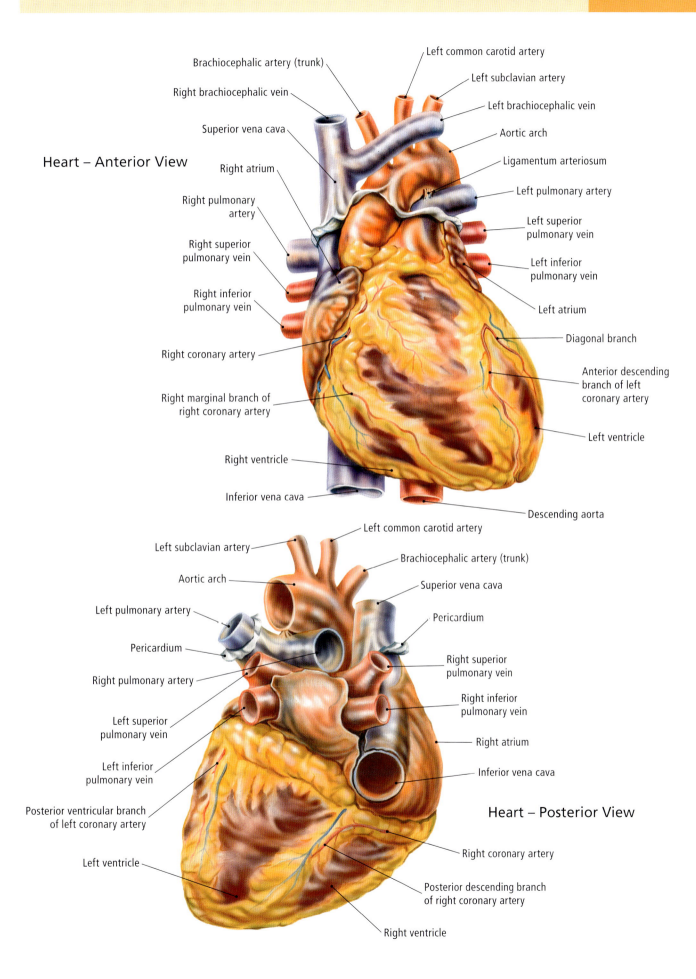

Heart – Anterior View

Brachiocephalic artery (trunk)
Right brachiocephalic vein
Superior vena cava
Right atrium
Right pulmonary artery
Right superior pulmonary vein
Right inferior pulmonary vein
Right coronary artery
Right marginal branch of right coronary artery
Right ventricle
Inferior vena cava

Left common carotid artery
Left subclavian artery
Left brachiocephalic vein
Aortic arch
Ligamentum arteriosum
Left pulmonary artery
Left superior pulmonary vein
Left inferior pulmonary vein
Left atrium
Diagonal branch
Anterior descending branch of left coronary artery
Left ventricle
Descending aorta

Left common carotid artery
Left subclavian artery
Aortic arch
Left pulmonary artery
Pericardium
Right pulmonary artery
Left superior pulmonary vein
Left inferior pulmonary vein
Posterior ventricular branch of left coronary artery
Left ventricle

Brachiocephalic artery (trunk)
Superior vena cava
Pericardium
Right superior pulmonary vein
Right inferior pulmonary vein
Right atrium
Inferior vena cava

Heart – Posterior View

Right coronary artery
Posterior descending branch of right coronary artery
Right ventricle

Blood Vessels of the Upper and Lower Limb

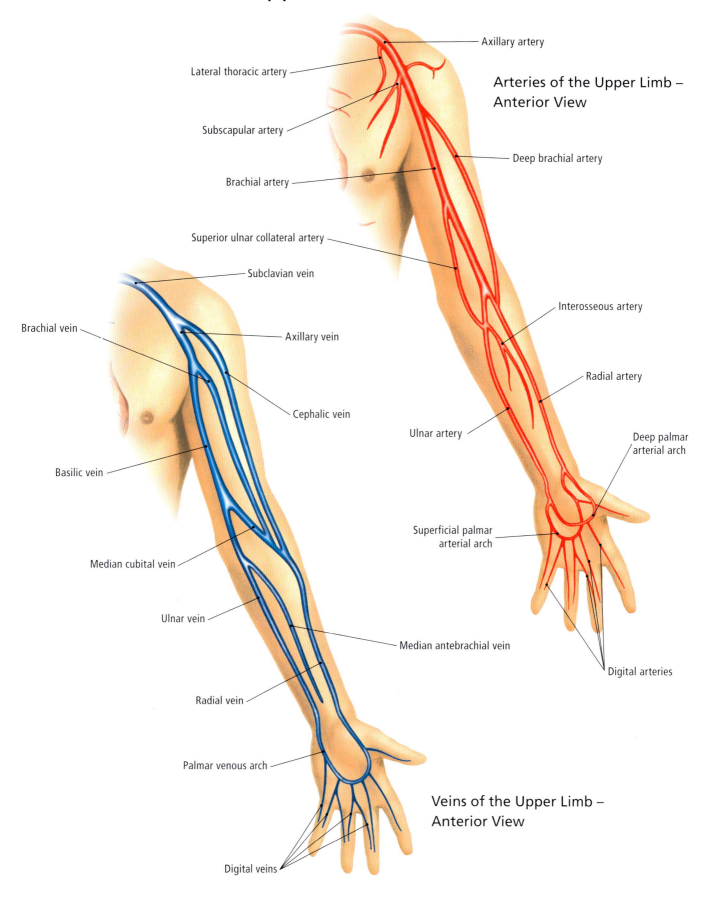

Axillary artery

Lateral thoracic artery

**Arteries of the Upper Limb –
Anterior View**

Subscapular artery

Deep brachial artery

Brachial artery

Superior ulnar collateral artery

Subclavian vein

Interosseous artery

Brachial vein

Axillary vein

Radial artery

Cephalic vein

Basilic vein

Ulnar artery

Deep palmar
arterial arch

Median cubital vein

Superficial palmar
arterial arch

Ulnar vein

Median antebrachial vein

Radial vein

Digital arteries

Palmar venous arch

**Veins of the Upper Limb –
Anterior View**

Digital veins

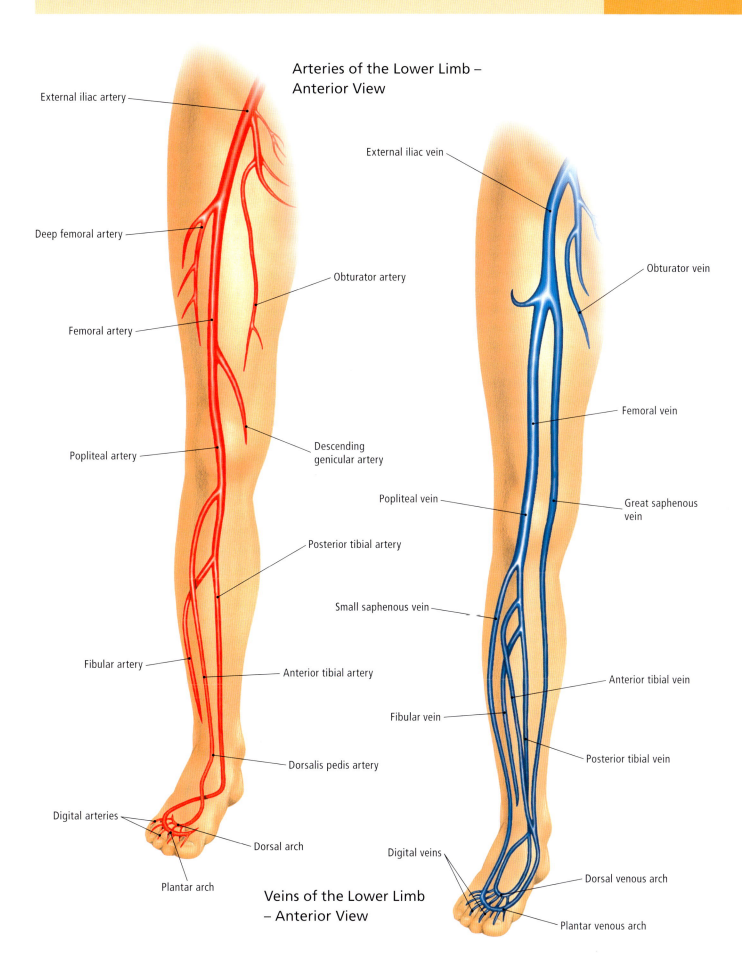

Arteries of the Lower Limb –
Anterior View

External iliac artery

Deep femoral artery

Obturator artery

Femoral artery

Popliteal artery

Descending
genicular artery

Posterior tibial artery

Fibular artery

Anterior tibial artery

Dorsalis pedis artery

Digital arteries

Dorsal arch

Plantar arch

Veins of the Lower Limb
– Anterior View

External iliac vein

Obturator vein

Femoral vein

Popliteal vein

Great saphenous
vein

Small saphenous vein

Anterior tibial vein

Fibular vein

Posterior tibial vein

Digital veins

Dorsal venous arch

Plantar venous arch

Respiratory System

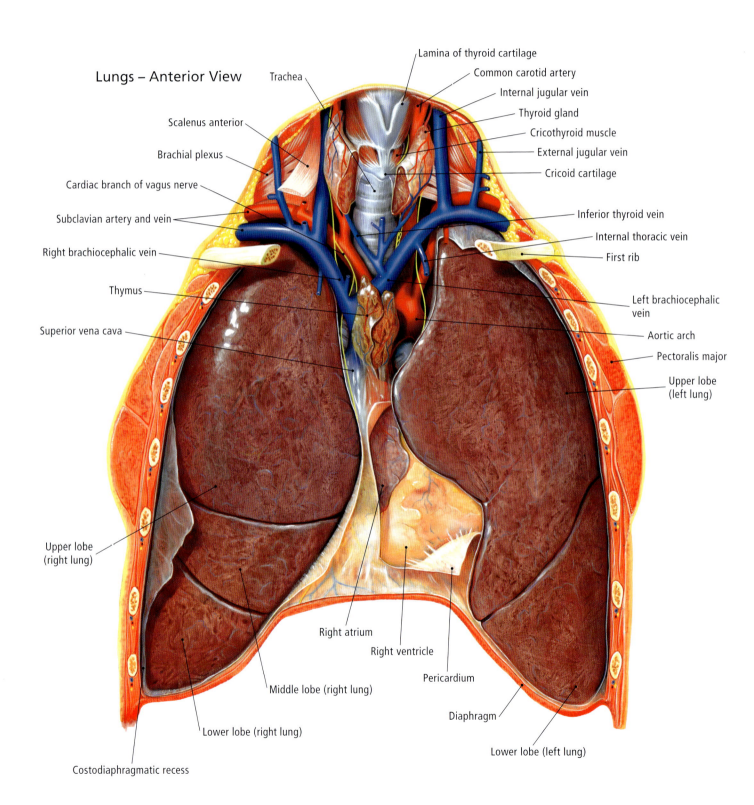

Lungs – Anterior View

Trachea

Lamina of thyroid cartilage

Common carotid artery

Internal jugular vein

Thyroid gland

Cricothyroid muscle

External jugular vein

Cricoid cartilage

Scalenus anterior

Brachial plexus

Cardiac branch of vagus nerve

Subclavian artery and vein

Right brachiocephalic vein

Thymus

Superior vena cava

Inferior thyroid vein

Internal thoracic vein

First rib

Left brachiocephalic vein

Aortic arch

Pectoralis major

Upper lobe (left lung)

Upper lobe (right lung)

Right atrium

Right ventricle

Pericardium

Diaphragm

Lower lobe (left lung)

Middle lobe (right lung)

Lower lobe (right lung)

Costodiaphragmatic recess

Respiratory System – Anterior View

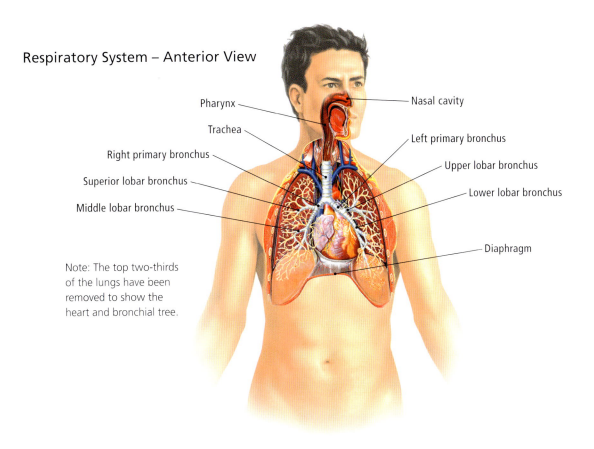

Pharynx

Nasal cavity

Trachea

Left primary bronchus

Right primary bronchus

Upper lobar bronchus

Superior lobar bronchus

Lower lobar bronchus

Middle lobar bronchus

Diaphragm

Note: The top two-thirds of the lungs have been removed to show the heart and bronchial tree.

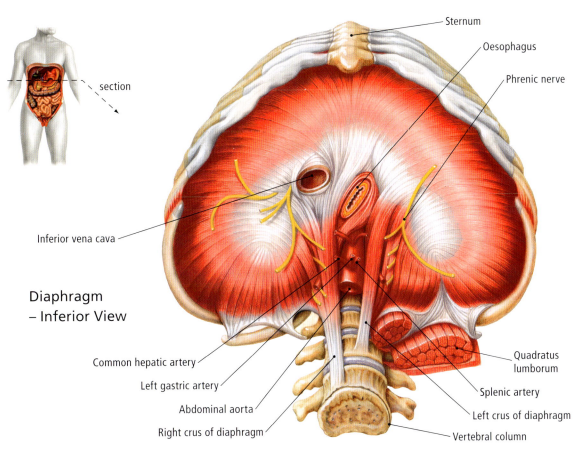

section

Sternum

Oesophagus

Phrenic nerve

Inferior vena cava

Diaphragm – Inferior View

Quadratus lumborum

Common hepatic artery

Splenic artery

Left gastric artery

Left crus of diaphragm

Abdominal aorta

Vertebral column

Right crus of diaphragm

Movements of the Body

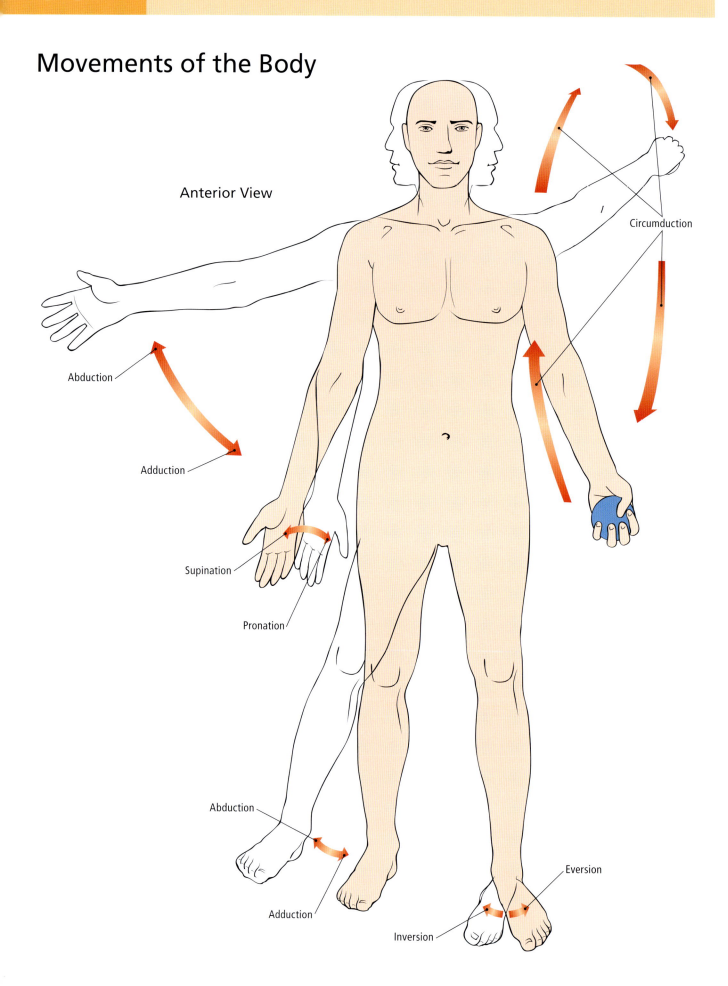

Anterior View

Circumduction

Abduction

Adduction

Supination

Pronation

Abduction

Adduction

Eversion

Inversion

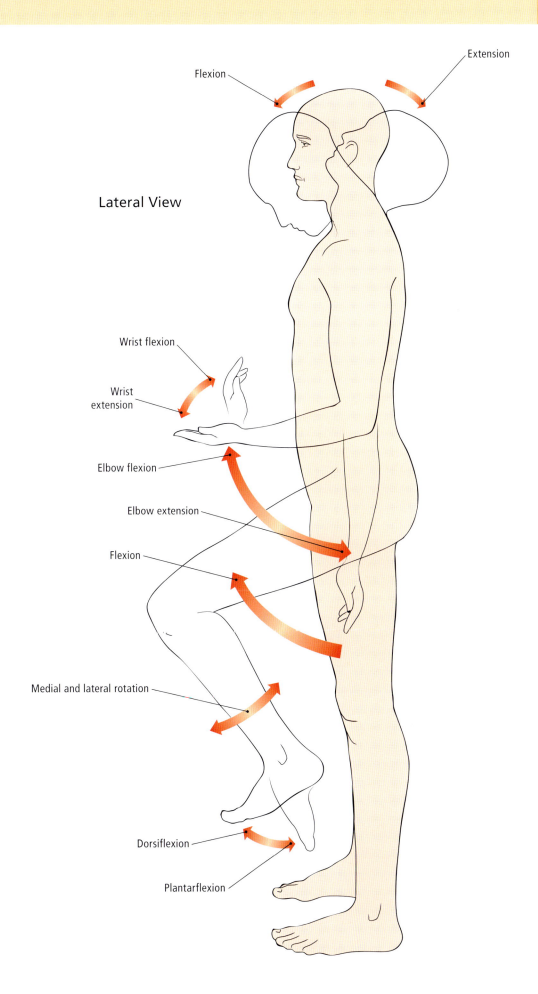

Flexion

Extension

Lateral View

Wrist flexion

Wrist extension

Elbow flexion

Elbow extension

Flexion

Medial and lateral rotation

Dorsiflexion

Plantarflexion

Exercises

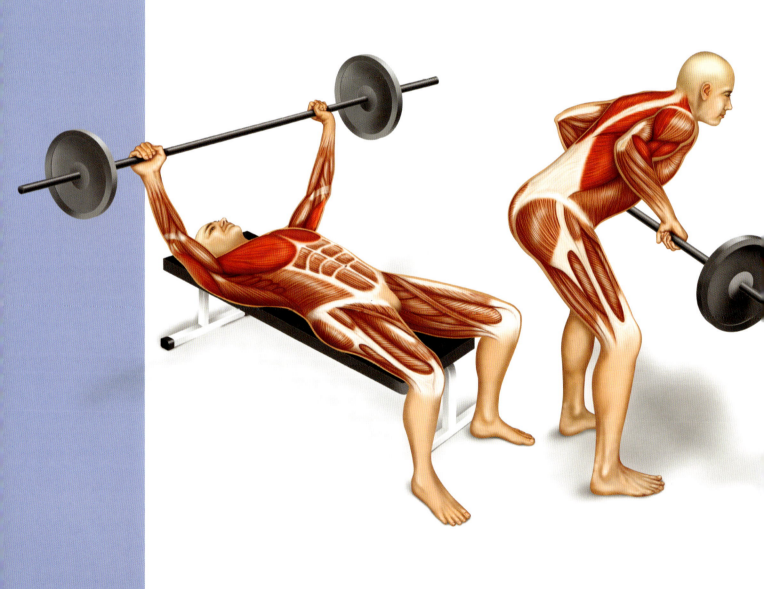

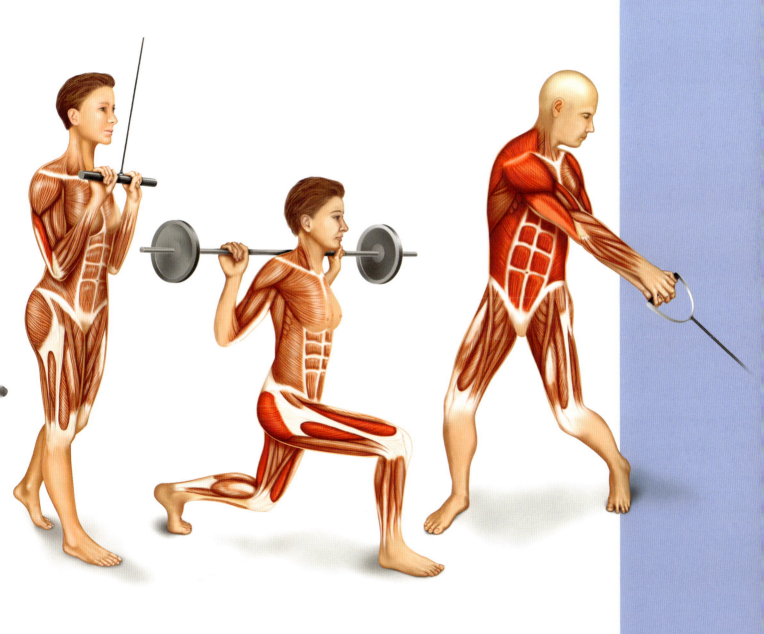

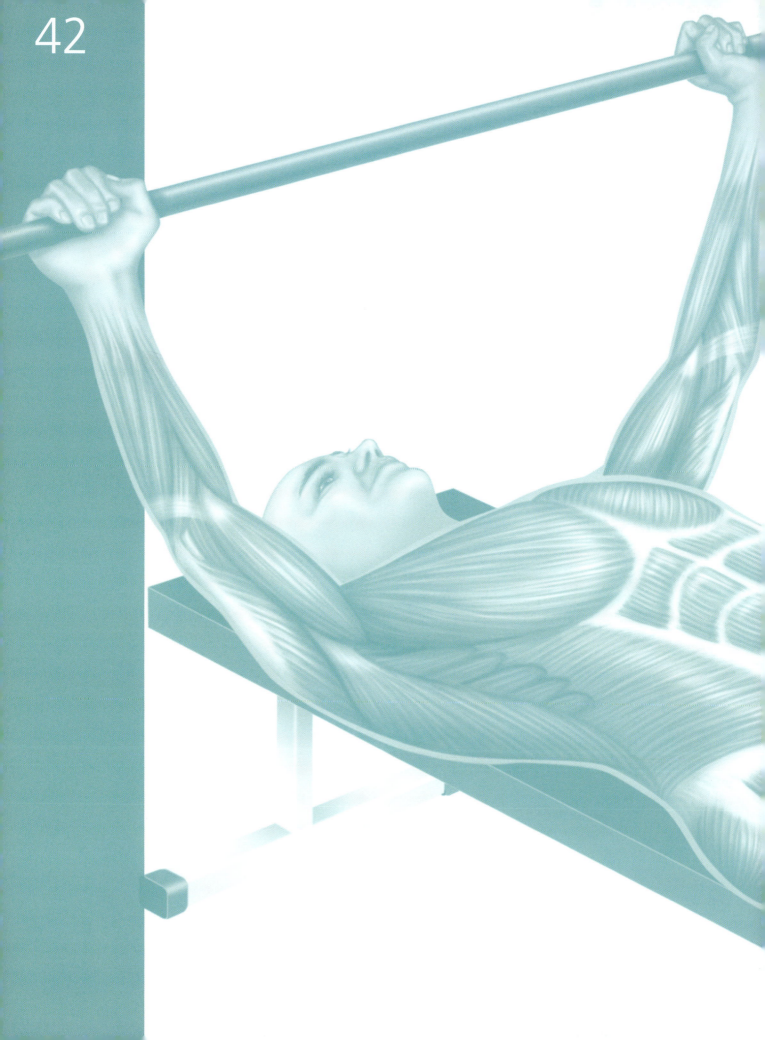

Chest Exercises

A strong chest improves posture, assists breathing, and helps protect the shoulders from injury. The main chest muscle is pectoralis major; other muscles include pectoralis minor, serratus anterior, and the intercostals. Pectoralis major has three primary actions: flexion, adduction, and internal (or medial) rotation of the arm. It is also an accessory muscle of breathing.

Most chest exercises involve pushing, and engage the triceps and deltoids as secondary muscles. Developing powerful pushing muscles makes many tasks easier, meaning less fatigue at the end of the day. For athletes, training chest muscles can help achieve longer and stronger throws, and improve their ability to push off opponents, grapple, and wrestle.

Dumbbell Chest Press

This exercise uses dumbbells to increase the demand on the shoulder and scapular stabilisers. It complements the bench press, as it requires both arms to work independently so that one side of the body cannot 'cheat' and let the other side do extra work. The dumbbell chest press targets the pectoral, deltoid, and triceps muscles, with a large stability component required from the rotator cuff, serratus anterior, rhomboids, trapezius, and latissimus dorsi muscles. This exercise helps improve performance in many everyday tasks, such as lifting and pushing, and is ideal to include if training for contact sports, throwing sports, or gymnastics.

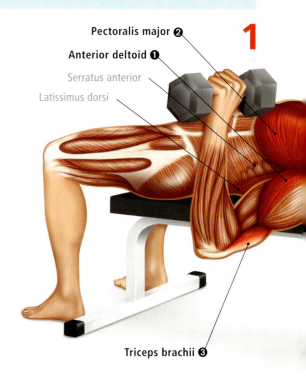

Pectoralis major ❷
Anterior deltoid ❶
Serratus anterior
Latissimus dorsi

1

Triceps brachii ❸

how to

To get into position, sit on the end of a bench with dumbbells resting on the knees. 'Kick' the weights up to the shoulders, then lie back flat on the bench. (The 'kick' is unnecessary if starting with light weights.) Hold the dumbbells at the sides of the chest, with palms facing down towards the body and elbows bent. Push up and extend the elbows following a small arc of motion so that the dumbbells come close together above the chest. Slowly lower weights until a slight stretch is felt in the front of the shoulder, then repeat.

variations

EASY Limit the range of motion in this exercise and do not lower weights past 90 degrees at the elbow if there is a history of shoulder injury. This shorter range of motion decreases the role of the shoulder stabilisers, but still allows heavy weights to be used.

HARD Use an inclined bench instead of a flat bench and reduce the amount of weight being lifted. This set-up still gives the chest muscles a great workout, but the change in bench angle increases the workload of the shoulder and triceps muscles at the same time.

active muscles

❶ Anterior deltoid
❷ Pectoralis major
❸ Triceps brachii

warning

Do not drop the weights from a lying position when finished, as this can lead to a shoulder dislocation.

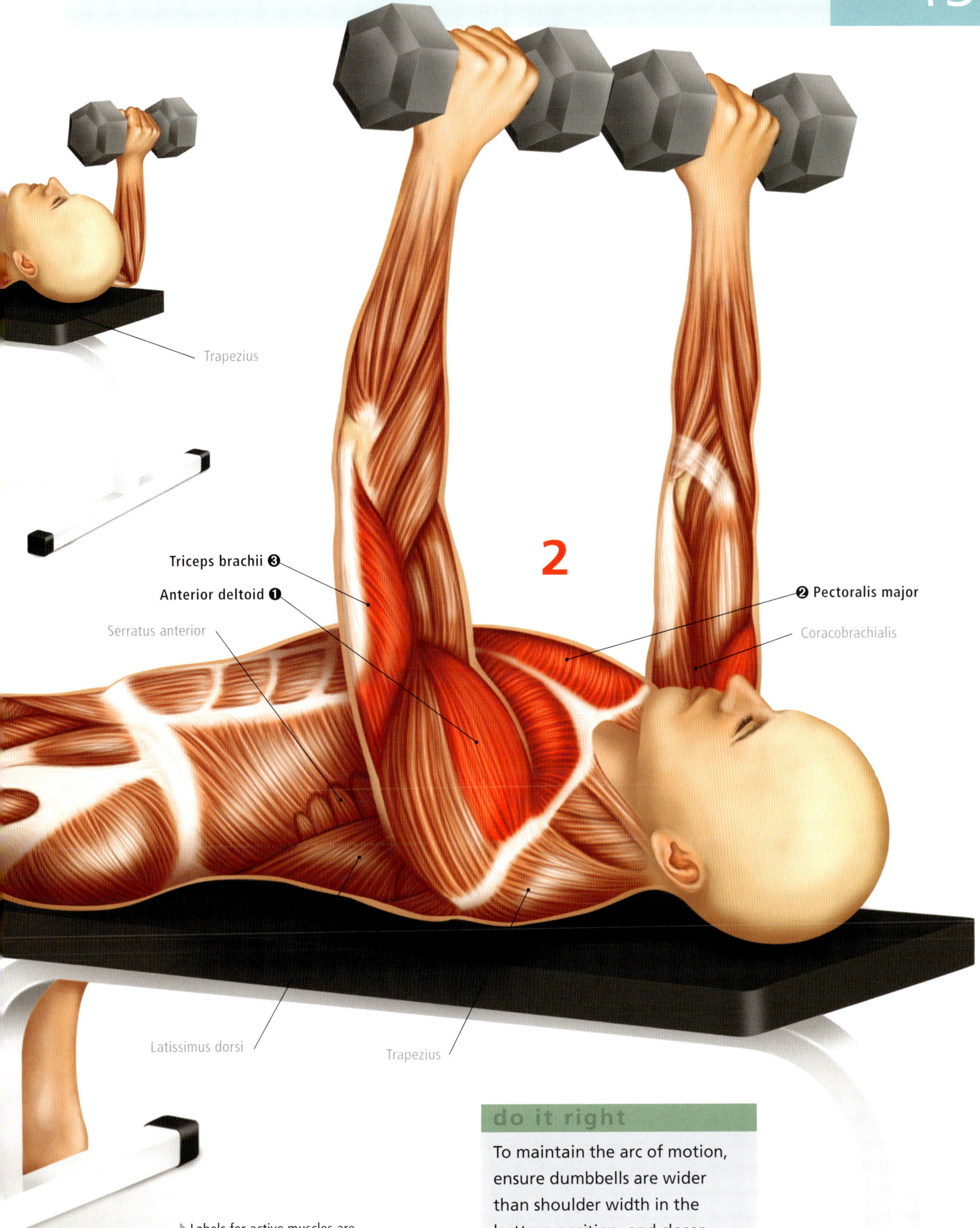
Trapezius
2
Triceps brachii ❸
Anterior deltoid ❶
Serratus anterior
❷ Pectoralis major
Coracobrachialis
Latissimus dorsi
Trapezius

▸ Labels for active muscles are shown in black, labels for stabiliser muscles are grey.

do it right

To maintain the arc of motion, ensure dumbbells are wider than shoulder width in the bottom position, and closer than shoulder width at the top.

Dumbbell Fly

This chest-building exercise requires the arm to move through an arc while the elbow remains at a constant angle. The pectoralis major muscle is the primary mover, with the deltoid called on to assist, and the elbow flexors – biceps brachii, brachioradialis, and brachialis – isometrically active. During the eccentric phase of the movement, the chest and arm muscles are also stretched. Because of the long lever arm, the amount of weight lifted to perform the dumbbell fly is significantly less than an equivalent chest exercise, such as the bench press. While this exercise is not as sports specific as the bench press, it is useful as a late-stage rehabilitation exercise for shoulder or elbow injuries.

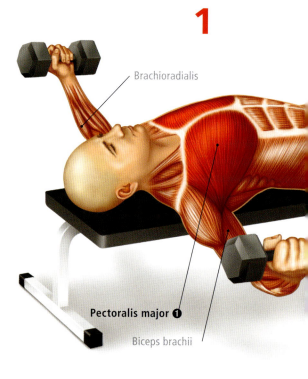

1

Brachioradialis

Pectoralis major ❶

Biceps brachii

how to

Lie back on a bench and start with hands in a neutral position, gripping the dumbbells above the middle of the chest. The elbows should remain slightly bent. Lower the dumbbells out to the sides of the chest, keeping the elbow angle constant, until a stretch is felt across the front of the shoulders and chest. Bring dumbbells back together at the mid-point above the chest.

variations

EASY

Lie on the floor instead of on a bench. This limits the range of motion and decreases the stretch on the shoulders. Muscles are weakest at their longest and shortest positions, so performing this exercise in the floor position also prevents the pectoralis major from stretching to the end of its range.

HARD

Try one-arm alternating dumbbell flys to isolate each side of the chest. Follow the same movement as for the standard dumbbell fly, but use one arm first, then alternate to the other. This variation provides an additional stability challenge, which can be increased again by performing the exercise lying on a stability ball, rather than on a bench.

active muscles

❶ Pectoralis major

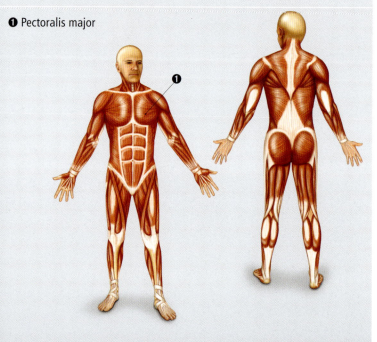

❶

warning

The end position is stressful and unstable for the shoulder joint. Do not risk injury by lifting weights that are too heavy.

do it right

Do not lock elbows. Maintain some elbow flexion throughout the movement but keep the elbow angle constant.

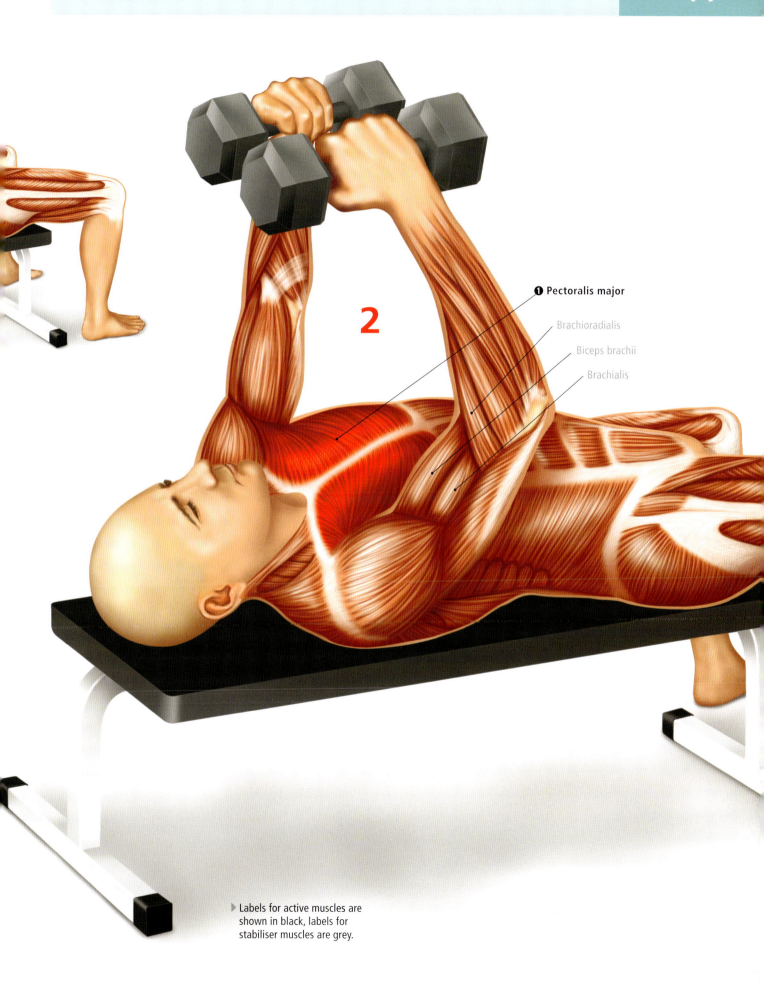

2

❶ **Pectoralis major**

Brachioradialis

Biceps brachii

Brachialis

▶ Labels for active muscles are
shown in black, labels for
stabiliser muscles are grey.

Bench Press

This classic compound exercise involves multiple joints and targets the chest, shoulder, and arm muscles. Heavy bench presses build size and strength in the pectoralis major, anterior deltoid, and triceps brachii, while also developing the rotator cuff muscles and scapular stabilisers. This exercise can be performed by both advanced and beginner weight trainers, because the basic movement is straightforward. It forms the basis of just about every upper-body strengthening programme, sports-specific conditioning programme, and advanced-rehabilitation programme.

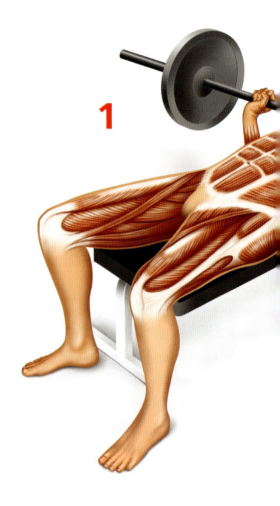

1

how to

Lie back on a flat bench and grasp the bar with both hands a little wider than shoulder-width apart, at an even distance from the weight at each end. Palms face up towards the ceiling. Take a deep breath in, then, while exhaling, lift the bar from the rack so that arms are fully extended and elbows locked. On the next breath in, slowly lower the bar towards the chest. Stop just short of touching the chest. Breathe out and push the weight straight up, away from the chest, returning to full elbow extension.

variations

EASY

Use just the barbell, without any additional weight, if learning the bench press. Focus on technique, and master smooth, controlled repetitions before loading up the bar.

HARD

Try a 'close grip' bench press. First, significantly reduce the weight from that of a normal bench press, as the close grip decreases the role of pectoralis major and the deltoid, to focus the work on your triceps muscle. Grasp the bar slightly closer than shoulder-width apart. Lower the weight towards the chest, keeping elbows close to the body. Ensure the grip is not too close, as this will destabilise the bar.

active muscles

❶ Pectoralis major
❷ Anterior deltoid
❸ Triceps brachii

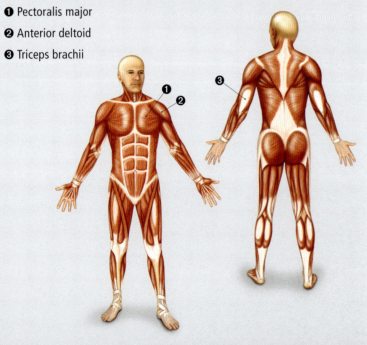

do it right

Do not arch the back with heavy lifts. This is a sign that the weight is too heavy and there is risk of injury.

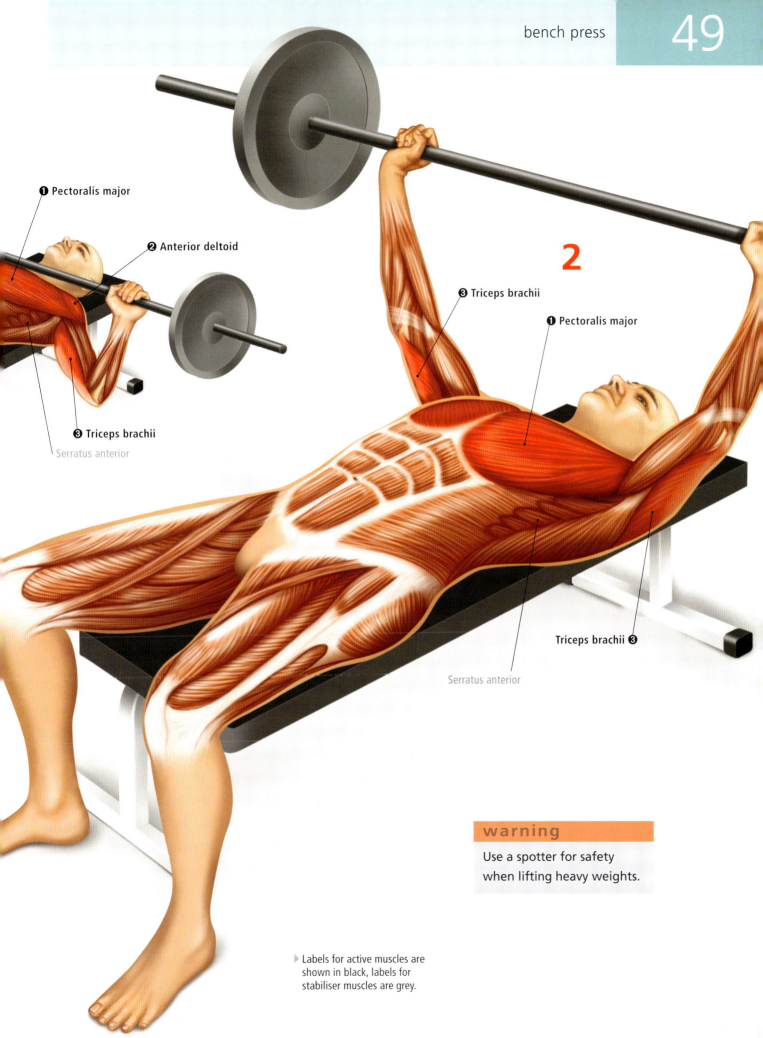

❶ Pectoralis major

❷ Anterior deltoid

❸ Triceps brachii

Serratus anterior

2

❸ Triceps brachii

❶ Pectoralis major

Triceps brachii ❸

Serratus anterior

▷ Labels for active muscles are shown in black, labels for stabiliser muscles are grey.

Dip

This bodyweight exercise complements the pull-up or chin-up, as far as upper body bodyweight exercises are concerned. The dip is a pushing exercise and, as such, it primarily targets the pectoralis major, triceps, and anterior deltoids. Alternatively, it may be possible to vary the dip to proportionally focus more on the triceps or pectoralis major, depending on the amount of forwards lean in the trunk. Performed on parallel dip bars, this exercise is difficult for beginners because it requires trainers to have enough strength to lift their own bodyweight, and strong abdominal muscles are needed to stabilise the core throughout the movement. Some gyms have an assisted-dip (and chin-up) machine that helps trainers who do not have the necessary strength to perform the dip unassisted.

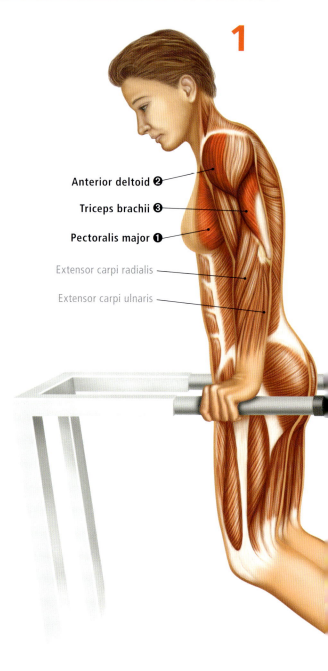

Anterior deltoid ❷

Triceps brachii ❸

Pectoralis major ❶

Extensor carpi radialis

Extensor carpi ulnaris

how to

Stand between the parallel dip bars and use the step so that the bars are at waist height. Grip the bars, lock the elbows, bend the knees, and lift the feet off the ground. Slowly lower the body using eccentric control of the extensor muscles, until elbows are bent to 90 degrees. Push back up until elbows are completely locked again.

variations

EASY

Bench dips are a good alternative if lacking upper body strength or access to an assisted-dip machine. Sit on a bench with legs extended out in front and heels together on the floor. Lift the upper body, supported by the arms. Shuffle forwards so that the body hovers in front of the bench, not above it. Slowly lower the body as per the standard dip, then push back up again until elbows are locked.

HARD

Increase the challenge by adding more weight. Use a dip belt, which enables the attachment of extra weight that hangs between the legs.

active muscles

❶ Pectoralis major

❷ Anterior deltoid

❸ Triceps brachii

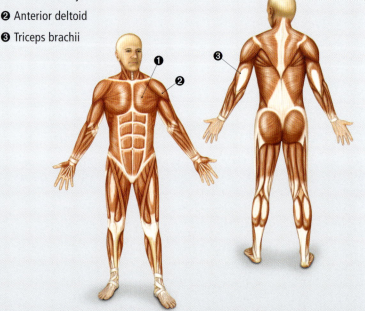

warning

Adding significant extra weight to the dip can cause pectoralis major and triceps tears. Do not add too much weight too quickly.

▶ Labels for active muscles are shown in black, labels for stabiliser muscles are grey.

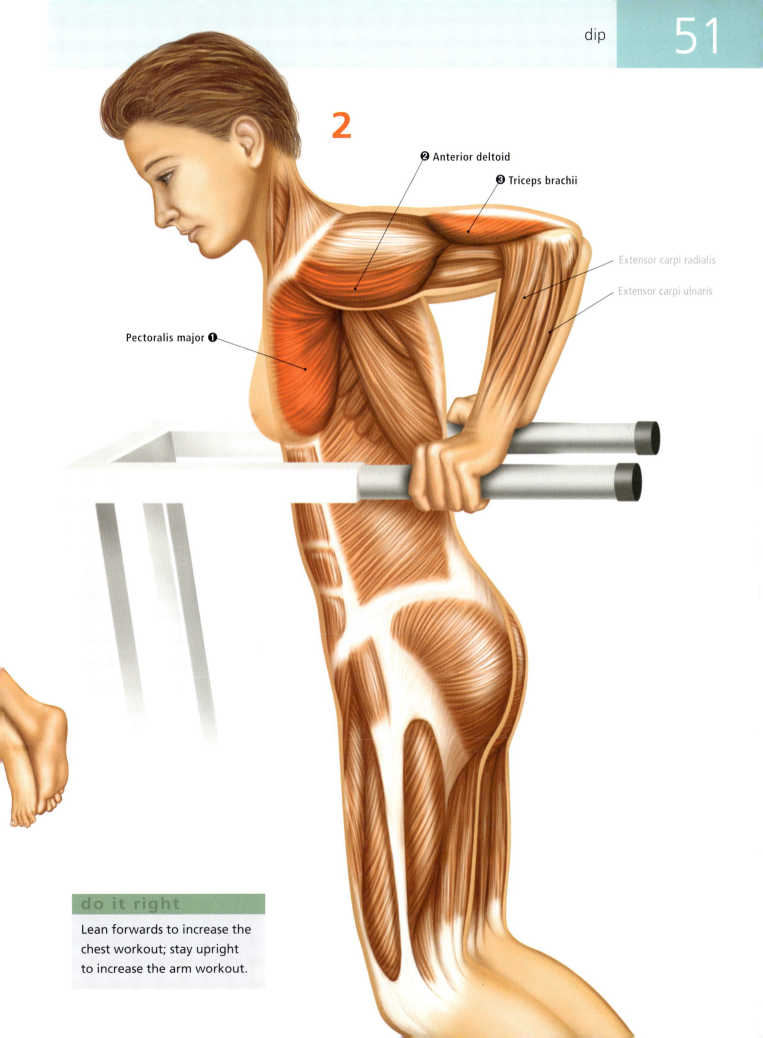

2

❷ Anterior deltoid

❸ Triceps brachii

Extensor carpi radialis

Extensor carpi ulnaris

Pectoralis major ❶

do it right

Lean forwards to increase the chest workout; stay upright to increase the arm workout.

Cable Crossover

This exercise, also known as the cable fly, isolates the chest muscles. The concentric phase of the movement helps strengthen the chest, while the eccentric phase provides a good stretch for both the chest and shoulders. Many trainers perform super set cable crossovers with other chest exercises, such as the bench press or dumbbell fly. Use this exercise as an alternative to the dumbbell fly to add variety to routines. The cable crossover is a good strengthening exercise for throwing sports, because of its unilateral nature and its similarity to the throwing motion. Adjust cable height and body position to feel different parts of the muscle working, or to make the movement more sport specific.

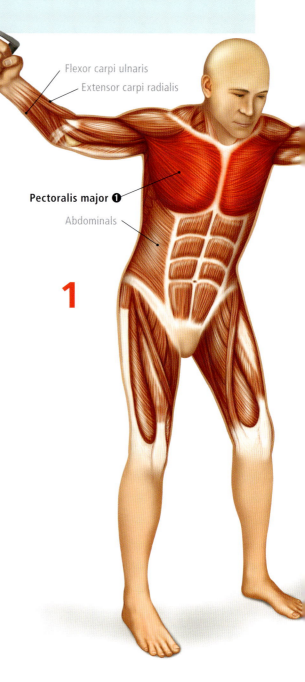

Flexor carpi ulnaris
Extensor carpi radialis

Pectoralis major ❶

Abdominals

1

how to

Stand in the middle of a cable-crossover machine with the pulleys adjusted to above-head height, equal on both sides. Grasp handles with palms facing down, shoulders internally rotated, and hips bent slightly forwards. Squeeze the chest muscles to pull arms downwards and inwards in a hugging motion. Keep elbows at a constant angle throughout the movement. Return to the start position with a slow, controlled motion.

variations

EASY

Lower pulleys to just below shoulder height and perform the standard exercise as described above. If the weight is too heavy, then technique will suffer and the effectiveness of the exercise will be lost. Use light to moderate weights and focus on form at all times.

HARD

Mid- and high-cable crossovers provide an all-around chest workout for advanced weight trainers. For mid crossovers, stand up straight and horizontally flex the arms to meet together in front of the chest. For high crossovers, start below the shoulders and bring hands together above the head.

active muscles

❶ Pectoralis major

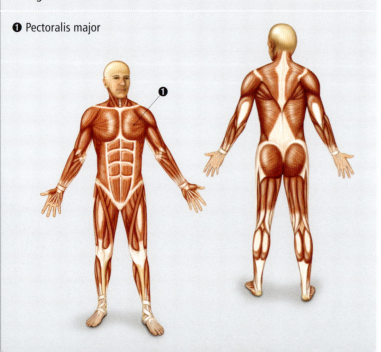

❶

warning

Do not let the arms be pulled back too quickly, as this can cause shoulder dislocations.

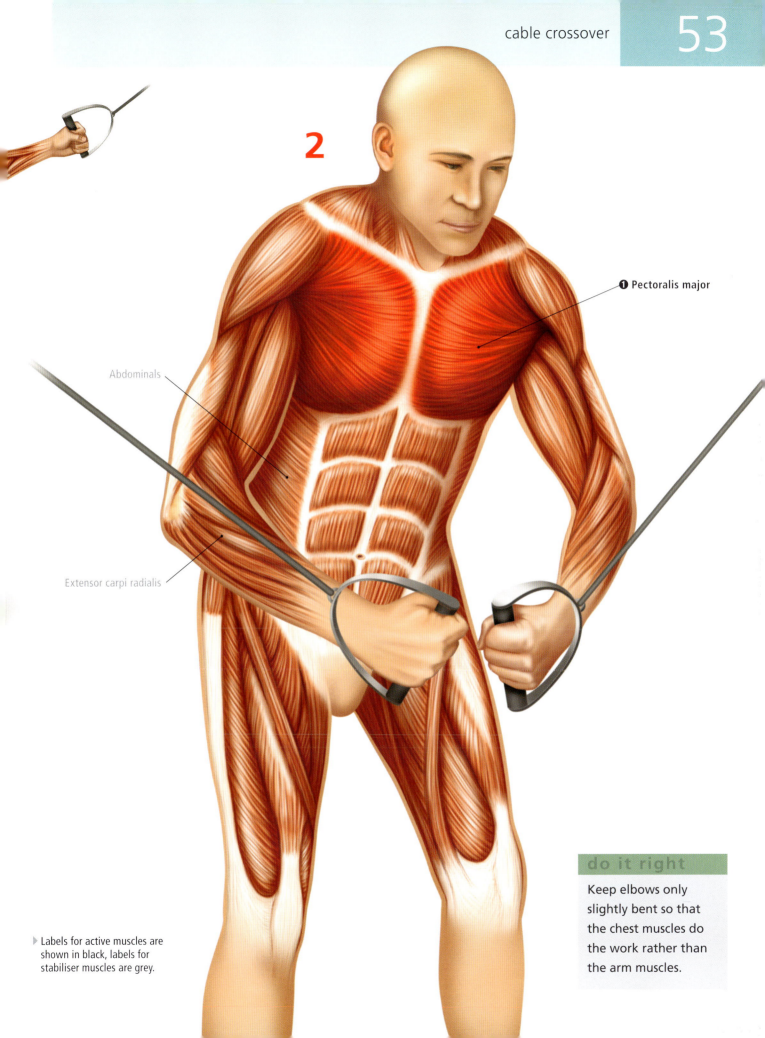

2

❶ Pectoralis major

Abdominals

Extensor carpi radialis

▶ Labels for active muscles are
shown in black, labels for
stabiliser muscles are grey.

do it right

Keep elbows only
slightly bent so that
the chest muscles do
the work rather than
the arm muscles.

Pullover

This chest and back exercise creates a great stretch for the pectoral muscles, the lateral muscles, and the abdominal muscles. Beginners may initially find the pullover difficult and intimidating, as the movement requires trainers to lift a dumbbell over the face. For this reason, it is best to start off with a light weight. Slowly build up the weight once completely comfortable with the technique. The pullover is an excellent exercise for throwing sports, as it develops strength and power. The stretching component will improve posture in anyone who has tight shoulders or an increased thoracic kyphosis – office workers typically fit this description.

warning

Range of motion varies according to individual flexibility; do not overstretch.

do it right

Keep elbows slightly bent and in line with the shoulders throughout the movement.

how to

Sit on a bench and grasp a dumbbell at one end with hands together, palms facing up. Lie back carefully, keeping the lower back in a neutral position. Push arms out until fully extended, holding the weight directly above the face. Slowly take the dumbbell back over the head. Keep abdominals tight to maintain a neutral spine. Reach back until a stretch is felt in the chest and shoulders. Keeping the elbows straight, pull the weight up to the starting position.

variations

EASY

Many gyms have a pullover machine that allows the same movement to be performed as the standard exercise, but without the need for core control, or the injury risk of dropping the dumbbell.

HARD

Try a standing cable pulldown. This variation increases the demand on core stability and works the posterior shoulder and the latissimus dorsi muscles. Maintaining a neutral spine while performing standing pulldowns is a significant challenge.

active muscles

❶ Pectoralis major
❷ Latissimus dorsi

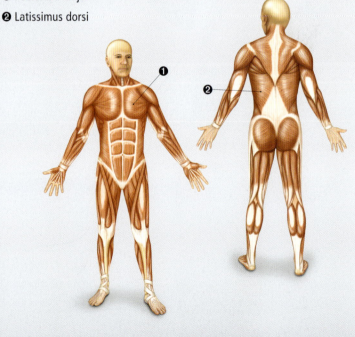

2

▸ Labels for active muscles are shown in black, labels for stabiliser muscles are grey.

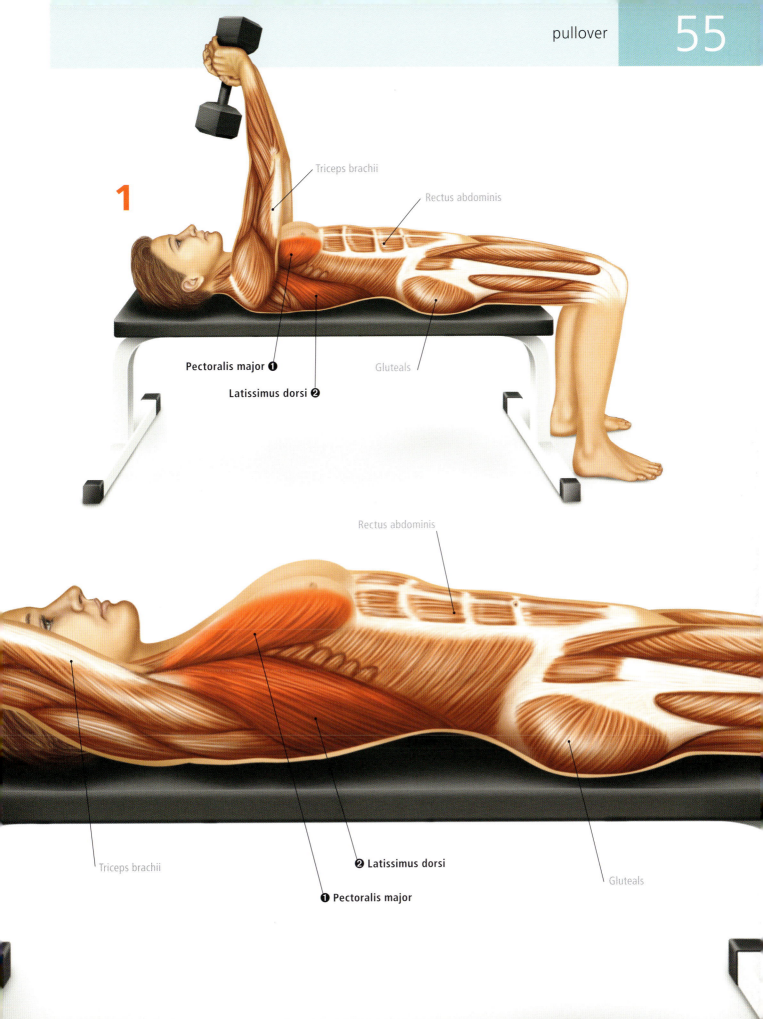

1

Triceps brachii

Rectus abdominis

Pectoralis major ❶

Latissimus dorsi ❷

Gluteals

Rectus abdominis

Triceps brachii

❷ Latissimus dorsi

❶ Pectoralis major

Gluteals

Push-up

This classic exercise is highly effective for improving the strength of the entire body. While the push-up primarily targets muscles in the chest, arms, and shoulders, it also requires support from other muscles. Because a wide range of muscles are integrated into the exercise, the push-up builds both upper body and core strength. It benefits the abdominal muscles by simultaneously flexing and stretching them. When the lower back muscles contract to stabilise the form, the abdominal muscles are inadvertently stretched. The quadriceps are also relied on to maintain proper form, giving the legs a secondary workout. Include the push-up in routines to stabilise the shoulders, as it develops both the scapular and rotator cuff muscles. This exercise does not require any equipment, so the push-up is well suited to daily maintenance routines.

1

how to

To start, lie face down on the ground with hands beside shoulders, fingers parallel to the body, and feet on their toes. Straighten arms, lifting the body and legs off the ground. Return to the starting position by bending arms and gently lowering the body until it hovers just above the ground.

variations

EASY

Place knees on the floor in the starting position if lacking a high level of upper body strength. Create a plane from the head to the knees as the push-up is performed. Ensure the body does not bend at the hips, as this causes the exercise to lose its effectiveness.

HARD

Place hands together under the body to focus on the triceps, or place them further away from the shoulders to target the chest muscles. While performing a series of basic push-ups, raise each leg in turn to work the lower back and gluteal muscles.

active muscles

❶ Anterior deltoid
❷ Pectoralis major
❸ Serratus anterior
❹ Triceps brachii

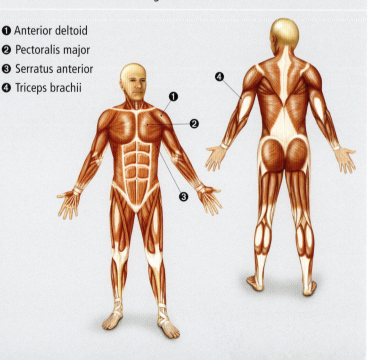

do it right

When lifting into the push-up position, keep body in a flat plane from head to ankles.

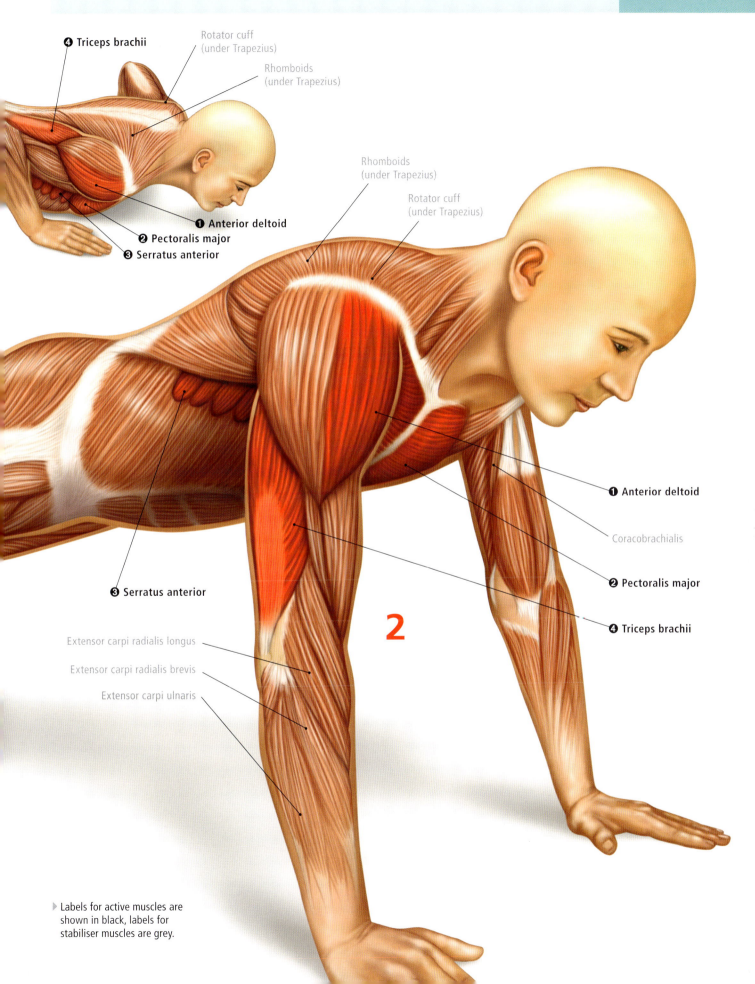

❹ Triceps brachii

Rotator cuff
(under Trapezius)

Rhomboids
(under Trapezius)

❶ Anterior deltoid

❷ Pectoralis major

❸ Serratus anterior

Rhomboids
(under Trapezius)

Rotator cuff
(under Trapezius)

❶ Anterior deltoid

Coracobrachialis

❷ Pectoralis major

❹ Triceps brachii

❸ Serratus anterior

Extensor carpi radialis longus

Extensor carpi radialis brevis

Extensor carpi ulnaris

2

▶ Labels for active muscles are
shown in black, labels for
stabiliser muscles are grey.

Back Exercises

As the back muscles are not seen in the mirror, they can be neglected during workouts in favour of chest, shoulder, and arm exercises. Do not make this mistake. Back muscles generate the force for pulling and lifting movements, support and protect the spine, and help control the scapula. Training these muscles assists in the performance of other exercises and a number of sporting activities, improves posture, and helps prevent injuries caused by unbalanced training programmes.

Choose an exercise combination that targets both the upper and lower back. For beginners, the goal is to learn the correct technique using basic lifts. For those with a developed technique and strength base, increase the variety, volume, and intensity of the weight-training regime.

Lat Pulldown

This well-known exercise is ideal for improving strength in both beginners and advanced exercisers. It targets the muscles of the back, shoulders, and arms, particularly the latissimus dorsi and the biceps, while the hip flexors and abdominal muscles work to stabilise the exerciser on the bench. Other back muscles are also called on to control and retract the scapula. The lat pulldown may improve posture and is recommended for sports that involve gripping and pulling movements. This is a gym-based exercise and should only be performed on the appropriate machine.

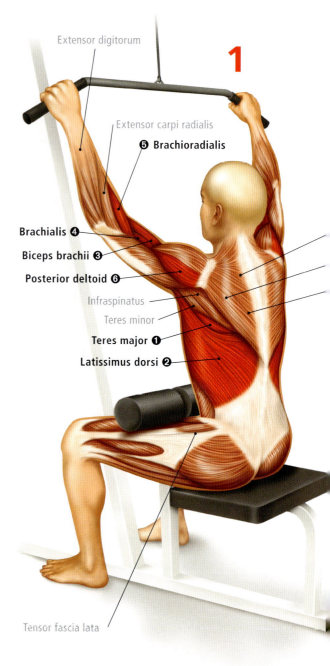

1

Extensor digitorum

Extensor carpi radialis

❺ Brachioradialis

Brachialis ❹

Biceps brachii ❸

Posterior deltoid ❻

Infraspinatus

Teres minor

Teres major ❶

Latissimus dorsi ❷

Tensor fascia lata

how to	Place hands on the bar just wider than shoulder-width apart, palms facing away from the body. Sit upright, with thighs firmly secured under the padding. Pull the bar down until it is under the chin, squeezing shoulder blades together. Lower the weight in one smooth, controlled motion until arms are straight.
variations — EASY	Try a 'close grip' lat pulldown if having trouble keeping a grip on the bar. Grasp the bar with hands just narrower than shoulder-width apart and palms facing towards the body. Pull the bar down until it is under the chin, squeezing shoulder blades together. Lower the weight in one smooth, controlled motion until arms are straight.
HARD	Replace the bar attachment with a rope attachment. This gives the arms and the back a great workout, while really challenging grip.

active muscles

❶ Teres major
❷ Latissimus dorsi
❸ Biceps brachii
❹ Brachialis
❺ Brachioradialis
❻ Posterior deltoid

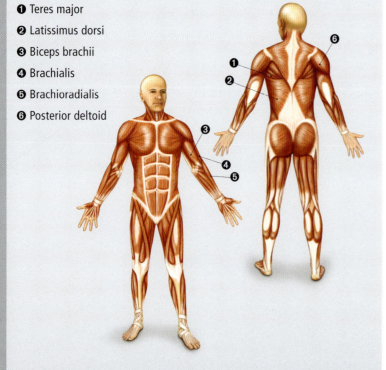

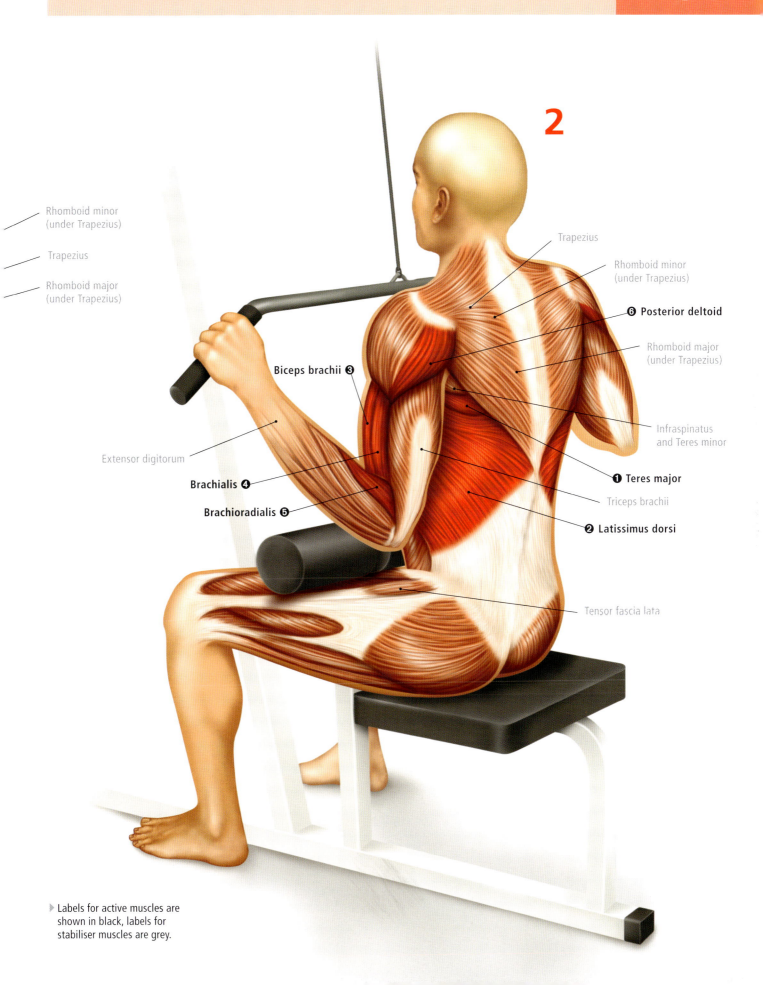

2

Rhomboid minor
(under Trapezius)

Trapezius

Rhomboid major
(under Trapezius)

Trapezius

Rhomboid minor
(under Trapezius)

❻ Posterior deltoid

Rhomboid major
(under Trapezius)

Infraspinatus
and Teres minor

Biceps brachii ❸

❶ Teres major

Extensor digitorum

Triceps brachii

Brachialis ❹

❷ Latissimus dorsi

Brachioradialis ❺

Tensor fascia lata

▶ Labels for active muscles are
shown in black, labels for
stabiliser muscles are grey.

Chin-up

This strength-building exercise is one that many people find difficult, as the exerciser must lift his or her own bodyweight. Chin-ups develop the back, shoulder, and arm muscles, particularly the latissimus dorsi and the biceps, and build grip strength in the fingers, hands, and forearms. Abdominal muscles are also given a good workout because of the stabilisation that is needed through the entire core. This exercise is recommended for any sport that involves gripping, grappling, and pulling, such as martial arts or rock climbing. It requires a stable bar in a gym or on an outdoor climbing frame, or can be performed on doorway bars installed in the home.

warning

Dropping back to the starting position suddenly can lead to hyperextension of the elbows and dislocation of the shoulder joints.

how to

Start with hands on the bar, a shoulder-width apart, with palms facing the body. Hang with knees slightly bent and head upright. In one smooth movement, pull body up until the chin is above the height of the bar, then gently lower body to the starting position. Ensure arms are fully extended when the chin-up is completed.

variations

EASY

Use a spotter to build strength if new to the exercise and finding it difficult to maintain correct technique. While in the starting position, bend at the knees so that the spotter can support the ankles. If necessary, push against this support base while raising body up towards the bar.

HARD

Try a pull-up to target the middle back muscles. The pull-up is performed in exactly the same way as the chin-up, except that hands are placed on the bar with palms facing away from the body. This technique is a favourite of the military and emergency services.

active muscles

❶ Trapezius
❷ Posterior deltoid
❸ Teres minor
❹ Teres major
❺ Biceps brachii
❻ Brachialis
❼ Brachioradialis
❽ Latissimus dorsi
❾ Rhomboid major
❿ Rhomboid minor

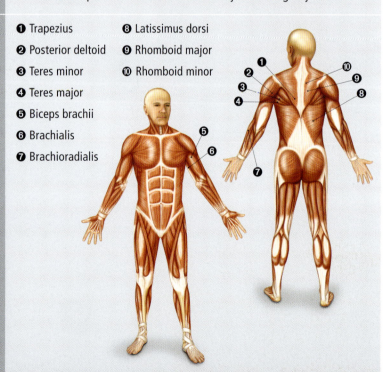

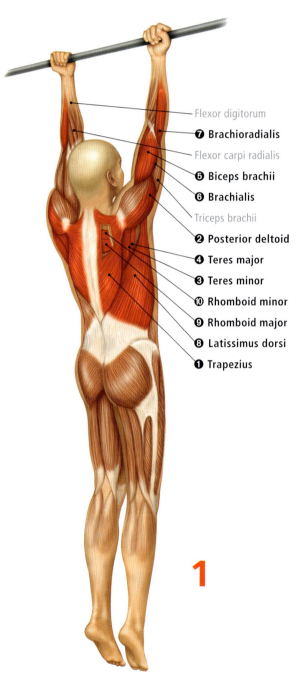

Flexor digitorum
❼ Brachioradialis
Flexor carpi radialis
❺ Biceps brachii
❻ Brachialis
Triceps brachii
❷ Posterior deltoid
❹ Teres major
❸ Teres minor
❿ Rhomboid minor
❾ Rhomboid major
❽ Latissimus dorsi
❶ Trapezius

1

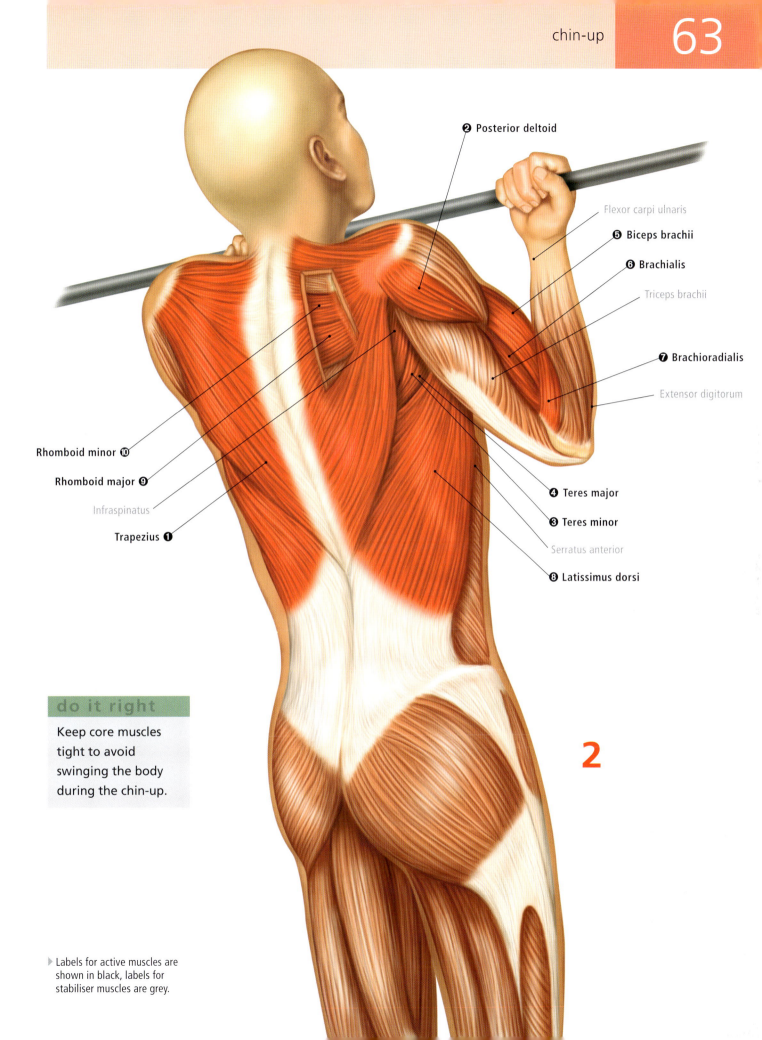

❷ Posterior deltoid

Flexor carpi ulnaris

❺ Biceps brachii

❻ Brachialis

Triceps brachii

❼ Brachioradialis

Extensor digitorum

❹ Teres major

❸ Teres minor

Serratus anterior

❽ Latissimus dorsi

Rhomboid minor ❿

Rhomboid major ❾

Infraspinatus

Trapezius ❶

do it right

Keep core muscles tight to avoid swinging the body during the chin-up.

2

▷ Labels for active muscles are shown in black, labels for stabiliser muscles are grey.

Bent-over Row

This free-weight exercise targets the muscles of the upper back, while calling on the lower back and leg muscles to provide support. In particular, the latissimus dorsi, rear deltoid, infraspinatus, and biceps are targeted, while the erector spinae and hamstring muscles must contract strongly to support the upper body. The bent-over row is suitable for intermediate or advanced exercisers and is ideal for sports and occupations that require bending, lifting, or pulling. It can be performed in either a gym or home setting, as it requires only some free floor space and a bar or dumbbells.

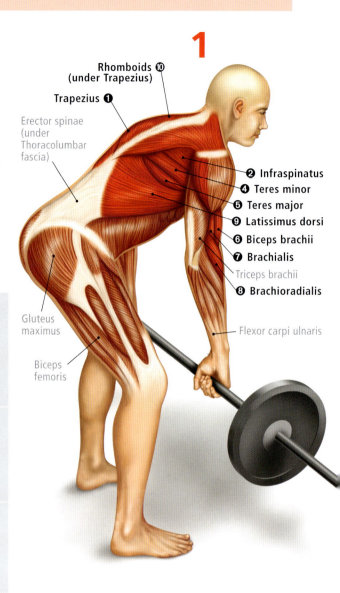

1

Rhomboids ⑩
(under Trapezius)

Trapezius ❶

Erector spinae
(under
Thoracolumbar
fascia)

❷ **Infraspinatus**
❹ **Teres minor**
❺ **Teres major**
❾ **Latissimus dorsi**
❻ **Biceps brachii**
❼ **Brachialis**
Triceps brachii
❽ **Brachioradialis**

Gluteus
maximus

Biceps
femoris

Flexor carpi ulnaris

how to	Grip the bar with hands just wider than shoulder-width apart, palms facing towards the body. Feet should be shoulder-width apart and knees slightly bent. In the starting position, bend forwards at the hips so the bar is just below the knees, with back straight. Pull bar up to the ribs, while keeping torso still and elbows close to sides. Lower again in a smooth, controlled movement.
variations — **EASY**	Use two dumbbells and grasp weights with palms facing each other. This makes the exercise easier on the grip and allows the weight to move, without the knees getting in the way. Pull the dumbbells up to the ribs, keeping elbows close to sides. Lower again in a smooth, controlled movement.
HARD	Perform the row with hands shoulder-width apart and palms facing away from the body to target the latissimus dorsi and provide an extra challenge for the biceps. Pull the bar up to the ribs, keeping elbows close to sides. Lower again slowly until arms are fully extended.

active muscles

❶ Trapezius
❷ Infraspinatus
❸ Posterior deltoid
❹ Teres minor
❺ Teres major
❻ Biceps brachii
❼ Brachialis
❽ Brachioradialis

❾ Latissimus dorsi
⑩ Rhomboids
(under Trapezius)

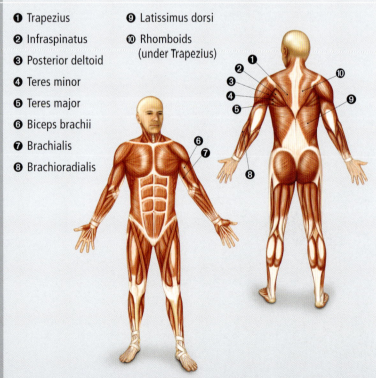

warning
Only bend forwards at the waist as far as flexibility allows. Do not round the back during this exercise.

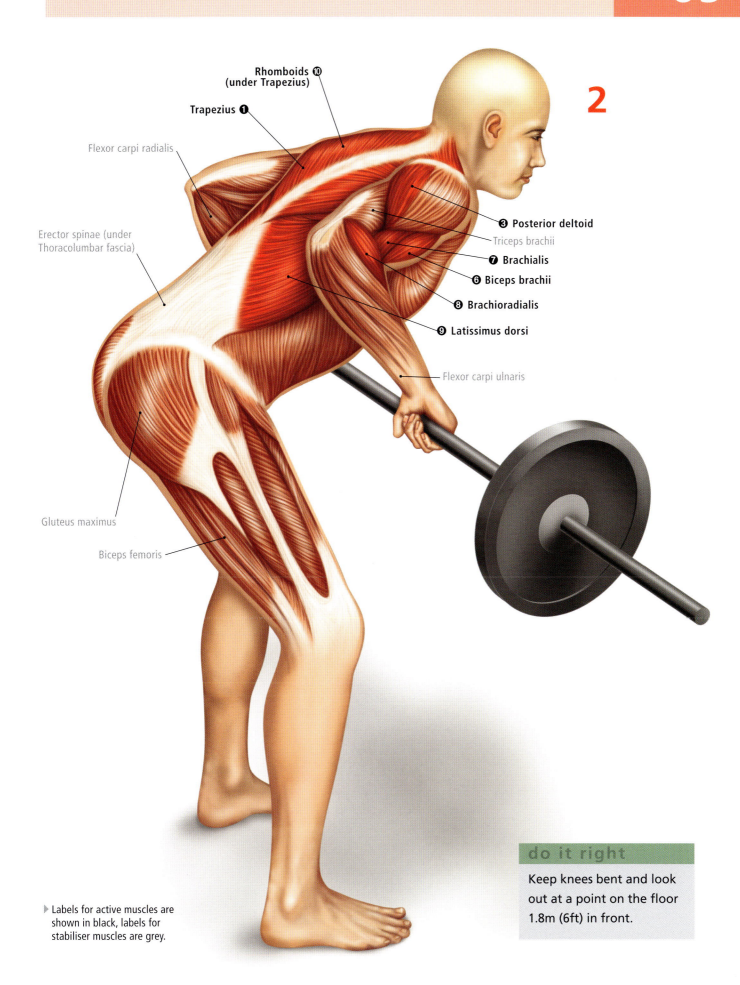

2

Rhomboids ❿
(under Trapezius)

Trapezius ❶

Flexor carpi radialis

Erector spinae (under
Thoracolumbar fascia)

❸ Posterior deltoid

Triceps brachii

❼ Brachialis

❻ Biceps brachii

❽ Brachioradialis

❾ Latissimus dorsi

Flexor carpi ulnaris

Gluteus maximus

Biceps femoris

do it right

Keep knees bent and look
out at a point on the floor
1.8m (6ft) in front.

▶ Labels for active muscles are
shown in black, labels for
stabiliser muscles are grey.

Seated Row

This strengthening exercise is performed in a gym setting. It is highly effective in providing a workout for the whole back. The muscles that control the scapula are given extra attention, with the latissimus dorsi, rear deltoids, and biceps bearing the brunt of the load. The leg, gluteal, and lower back muscles are also required to provide a stable support base, bracing in the sitting position. Use the seated row to add some postural balance back into a workout routine because the muscles that retract and depress the scapula are often weak and underutilised. This exercise requires an appropriate seated-row machine or a low cable pulley. It can be performed by both beginners and advanced trainers.

1

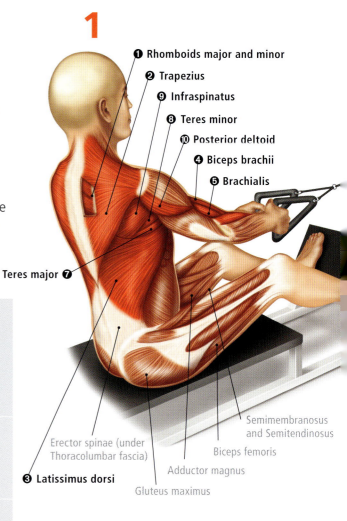

❶ Rhomboids major and minor
❷ Trapezius
❾ Infraspinatus
❽ Teres minor
❿ Posterior deltoid
❹ Biceps brachii
❺ Brachialis
Teres major ❼
Erector spinae (under Thoracolumbar fascia)
❸ Latissimus dorsi
Gluteus maximus
Adductor magnus
Biceps femoris
Semimembranosus and Semitendinosus

how to

Most machines have a handle where the hands are close together. Grasp this with palms facing towards each other. Place feet on the footholds and bend knees at about 30 degrees. Start with arms fully extended, then squeeze shoulder blades together and pull the bar towards the ribs, so elbows brush by sides. Allow the weight to extend slowly back to the starting position.

variations

EASY

Use a machine that has a chest support, which allows the exercise to be performed without placing any demand on the lower back or legs. However, progress to the standard version of the seated row once the easy variation has been mastered.

HARD

Use a long bar with hands just wider than shoulder-width apart, palms facing down. This reduces the load on the latissimus dorsi and targets the rear deltoids and rhomboid muscles. Squeeze shoulder blades together and pull the bar towards the chest, keeping elbows level with the bar.

active muscles

❶ Rhomboids major and minor
❷ Trapezius
❸ Latissimus dorsi
❹ Biceps brachii
❺ Brachialis
❻ Brachioradialis
❼ Teres major
❽ Teres minor
❾ Infraspinatus
❿ Posterior deltoid

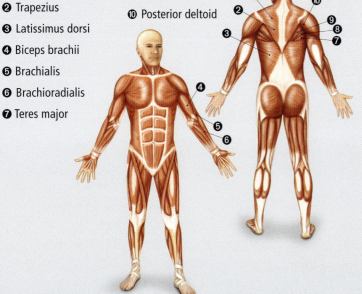

warning

Leaning forwards or backwards during the movement places unnecessary stress on the lower back.

2

do it right
Maintain a tall sitting position throughout the whole movement.

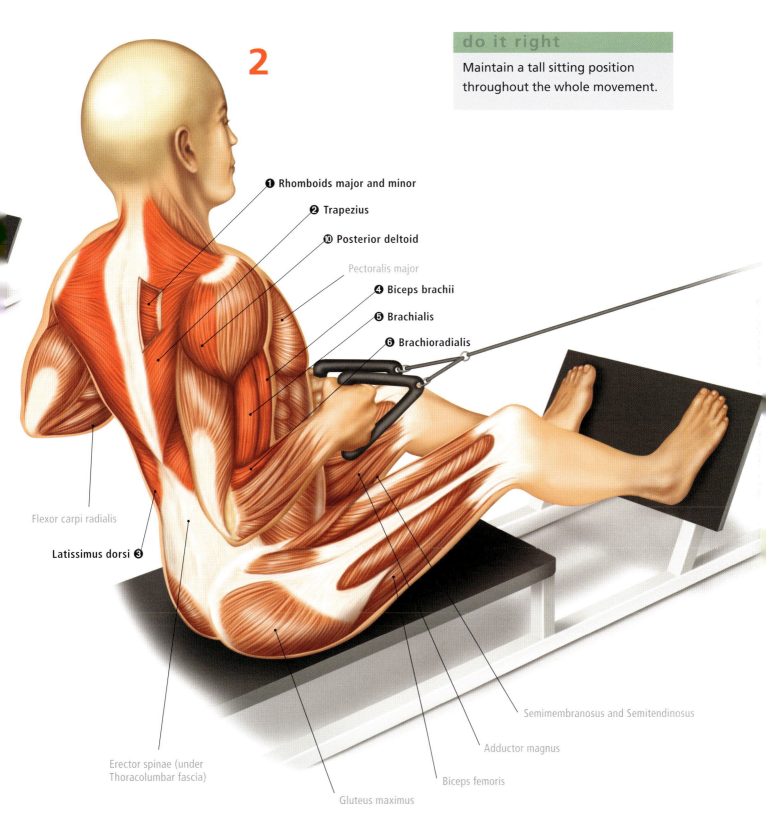

❶ Rhomboids major and minor

❷ Trapezius

❿ Posterior deltoid

Pectoralis major

❹ Biceps brachii

❺ Brachialis

❻ Brachioradialis

Flexor carpi radialis

Latissimus dorsi ❸

Semimembranosus and Semitendinosus

Adductor magnus

Erector spinae (under Thoracolumbar fascia)

Biceps femoris

Gluteus maximus

▶ Labels for active muscles are shown in black, labels for stabiliser muscles are grey.

Reverse Fly

This simple-but-effective exercise targets the muscles of the upper back and shoulders. The rear deltoid does most of the heavy lifting, assisted by the infraspinatus, teres minor, trapezius, and rhomboids. Recommended for both beginners and advanced exercisers, the reverse fly should be scheduled at the end of a workout, after the pulling exercises, to make sure these muscles have been worked hard enough during the session. This exercise can help prevent common shoulder injuries experienced in racket sports because it strengthens the muscles around the shoulder. Two dumbbells, a bit of space, and a sturdy bench or chair to sit on are all that is needed to perform the reverse fly.

warning

At the top of the movement, elbows should be at right angles to the torso to avoid placing stress on the rotator cuff muscles.

how to

Sit at the end of a chair or bench and lean forwards so the torso rests on the thighs. Grasp the dumbbells either underneath the legs or just next to the feet, with palms facing towards each other. Raise arms to the side, squeezing shoulder blades together, until elbows are shoulder height. Slowly lower back into the starting position, being careful that dumbbells do not collide with the ankles.

variations

EASY Lie face down on a bench to perform the reverse fly if unable to get into a comfortable sitting position. Ensure that the back does not arch during the lift.

HARD Perform the exercise in the standing position, with feet hip-width apart and knees slightly bent. Lean forwards at the hips, keeping the back straight. Replace dumbbells with a low pulley cable, or for an even greater challenge, exercise one arm at a time.

active muscles

❶ Posterior deltoid

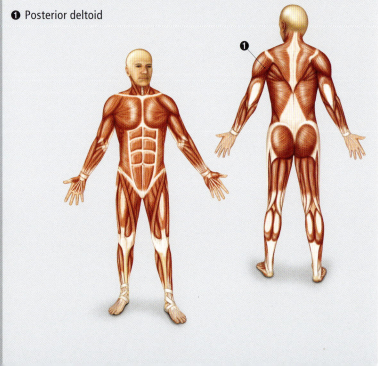

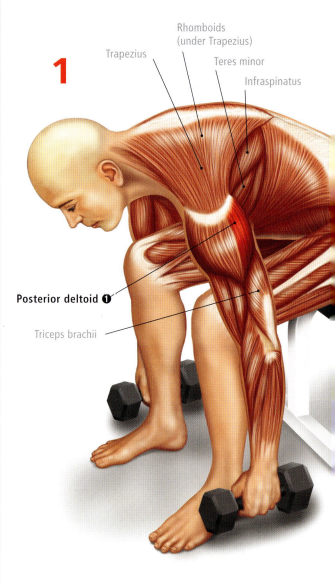

1

Trapezius

Rhomboids (under Trapezius)

Teres minor

Infraspinatus

Posterior deltoid ❶

Triceps brachii

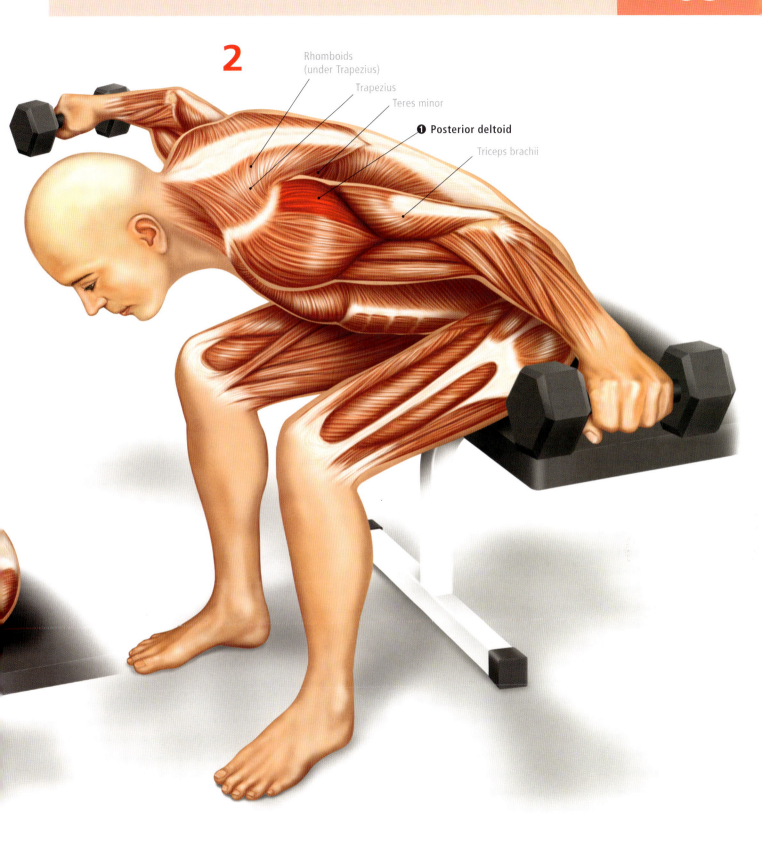

2

Rhomboids
(under Trapezius)

Trapezius

Teres minor

❶ **Posterior deltoid**

Triceps brachii

▷ Labels for active muscles are
shown in black, labels for
stabiliser muscles are grey.

do it right

Keep elbows slightly bent
throughout this exercise.

Single-arm Row

This exercise provides the back muscles with a great workout in a supported, stable position. The latissimus dorsi, rear deltoids, and biceps muscles are targeted, while the muscles around the scapula, forearm, and grip strength are also developed. The single-arm row is performed in a position that supports the lower back, so it suits both beginners and advanced exercisers. It is ideal for any sports that involve gripping and pulling, such as contact sports, rowing, and kayaking, and is a powerful exercise for increasing strength. All that is required is a sturdy bench at approximately knee height and one dumbbell.

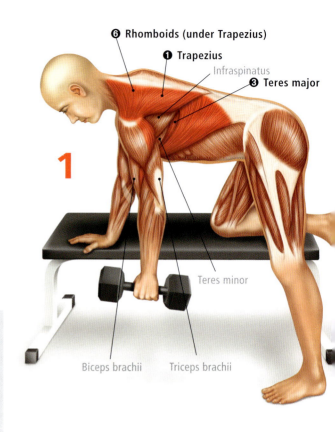

❻ Rhomboids (under Trapezius)
❶ Trapezius
Infraspinatus
❸ Teres major

1

Teres minor

Biceps brachii   Triceps brachii

how to

Place knee and hand of the supporting arm on a bench, so torso is horizontal. Position the other foot on the ground, slightly back and to the side for stability. Grasp the dumbbell with palm facing towards the bench. Pull the dumbbell up towards the body until it is at torso height, with elbow brushing by the side of the body. Slowly lower the weight until the arm is fully extended.

variations

EASY Lie face down on the bench if uncomfortable or feeling unstable in the kneeling position. Ensure the back does not arch during the lift.

HARD For an extra challenge, perform the exercise in a standing position, with feet shoulder-width apart and knees bent. Try using a low pulley cable without any support for the opposite hand.

active muscles

❶ Trapezius
❷ Posterior deltoid
❸ Teres major
❹ Brachialis
❺ Latissimus dorsi
❻ Rhomboids (under Trapezius)

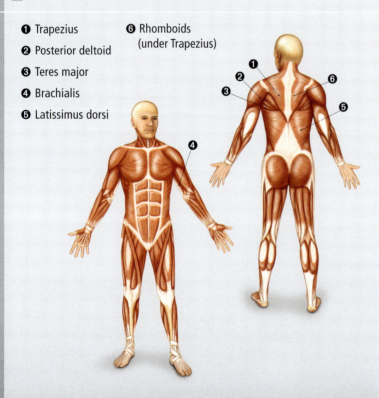

warning

Do not rotate torso to swing the weight up.

do it right

Keep back straight during the lift and look at a spot on the ground 1.2m (4ft) in front.

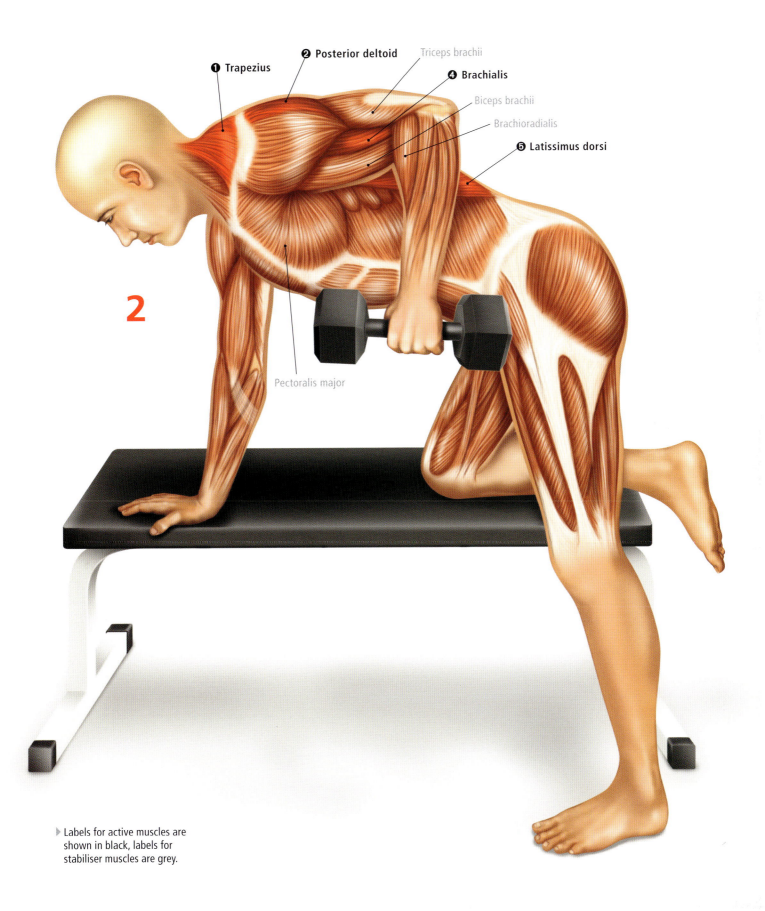

❶ Trapezius

❷ Posterior deltoid

Triceps brachii

❹ Brachialis

Biceps brachii

Brachioradialis

❺ Latissimus dorsi

2

Pectoralis major

▷ Labels for active muscles are shown in black, labels for stabiliser muscles are grey.

Arm and Shoulder Exercises

The deltoid is the target of shoulder exercises in this section. Its three sets of fibres generate force, while the rotator cuff muscles provide dynamic stability and allow for the range of motion around the shoulder (glenohumeral) joint. Strong, stable arms and shoulders enhance performance in many sports. Arm exercises address the muscles that flex the elbow – biceps brachii, brachialis, and brachioradialis – as well as triceps brachii, which extends the elbow.

Both shoulder and arm muscles are engaged during exercises for other muscle groups. To avoid fatigue and overuse injuries, allow adequate rest and recovery between workouts. The anterior deltoid and rotator cuff muscles are particularly susceptible, so it is important to pay attention to the posterior deltoid.

Biceps Curl

This classic exercise has been a staple of strength-training regimens for decades. It is a simple exercise that can be performed by both beginners and advanced trainers, and has many variations. The biceps curl targets the biceps brachii, but also gives other muscles of the upper arm, brachialis, and brachioradialis a good workout. The muscles of the shoulder and forearm must also contract to support the movement this exercise demands. Requiring only some floor space and a barbell or pair of dumbbells, this exercise can be performed almost anywhere.

1

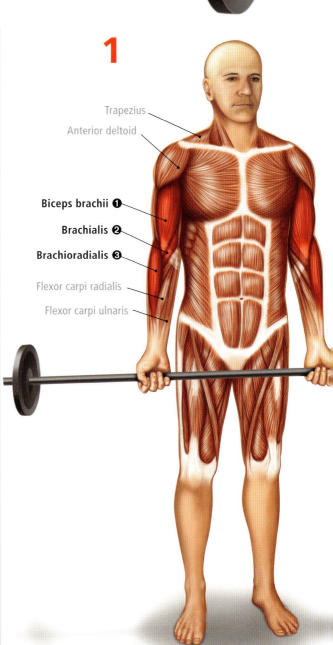

Trapezius
Anterior deltoid
Biceps brachii ❶
Brachialis ❷
Brachioradialis ❸
Flexor carpi radialis
Flexor carpi ulnaris

how to

Stand with feet shoulder-width apart and knees slightly bent. Grasp the bar with hands shoulder-width apart, palms facing away from the body. Move hands closer or further apart if these positions are more comfortable. Raise the bar until forearms are vertical, keeping elbows fixed by sides. Lower the bar until arms are fully extended.

variations

EASY

Stand with one foot forwards and one foot back for extra stability if feeling unsteady or lifting heavy loads. This position also helps keep the torso still and provides extra support for the lower back.

HARD

Sit or stand with one dumbbell in each hand. Start with palms facing towards each other, then with each dumbbell curl, rotate wrists so that palms face the shoulder at the top of the movement.

active muscles

❶ Biceps brachii
❷ Brachialis
❸ Brachioradialis

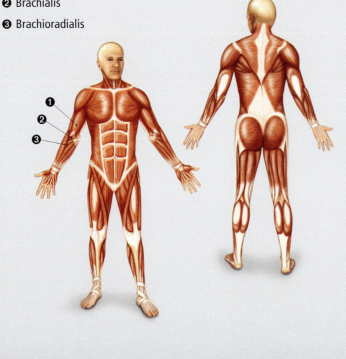

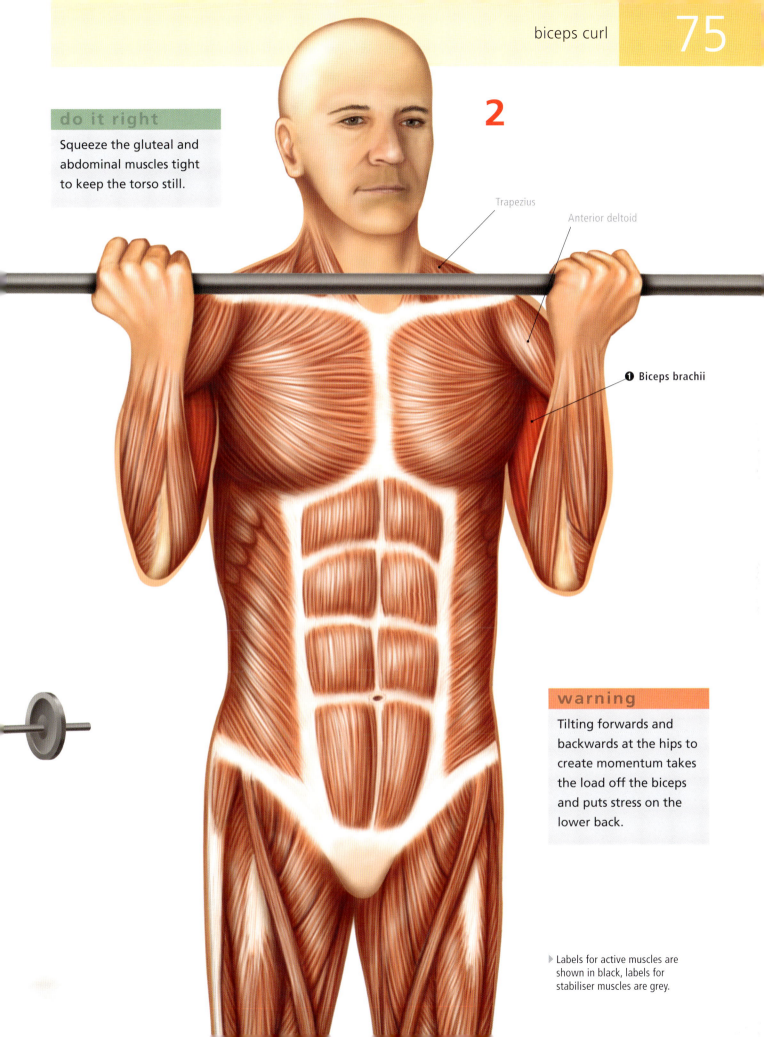

2

Squeeze the gluteal and abdominal muscles tight to keep the torso still.

Trapezius

Anterior deltoid

❶ Biceps brachii

warning

Tilting forwards and backwards at the hips to create momentum takes the load off the biceps and puts stress on the lower back.

▶ Labels for active muscles are shown in black, labels for stabiliser muscles are grey.

Concentration Curl

This simple exercise targets the muscles of the upper arm. Brachialis is the most active of these muscles, as shoulder flexion disadvantages the short head of biceps brachii. This focus makes the concentration curl a good addition, or at least a good occasional alternative, to more traditional biceps exercises. Include it in an exercise programme to ensure maximum upper arm workout. The stable seated position reduces stress on the lower back, limits the use of momentum, and allows the muscles of the upper arm to be isolated.

warning

Avoid twisting, rotating, or lifting the torso in order to lift the weight. If any of these actions are required, the weight is too heavy.

do it right

Place other hand on the other thigh to support the upper body.

how to

Sit on a bench with the dumbbell between the feet. Lean forwards slightly so that the elbow rests against the inner thigh. Grasp the dumbbell with palm facing away from the body, then curl the arm up until the palm faces the shoulder. Stabilise the arm with the thigh, so the elbow remains still throughout the exercise. Lower the weight until the arm is fully extended.

variations

EASY Place the elbow on top of the thigh rather than against the inner thigh, to allow for a little extra leverage and to make the exercise easier at the bottom of the lift.

HARD Perform a 'hammer curl' by rotating the wrist during the curl, so the thumb faces the ceiling at the top of the movement, with palm facing inwards. This variation targets the other elbow flexor muscle, brachioradialis.

active muscles

❶ Brachialis
❷ Biceps brachii
❸ Brachioradialis

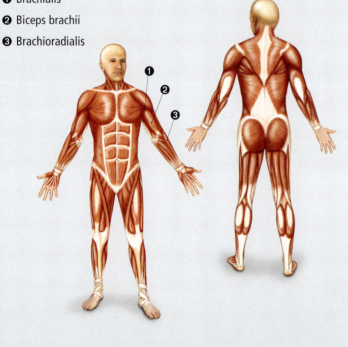

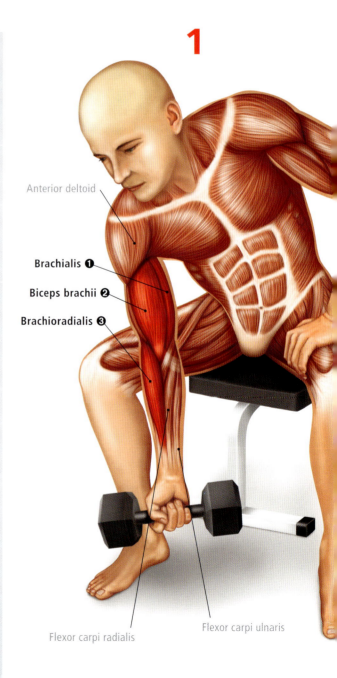

1

Anterior deltoid

Brachialis ❶

Biceps brachii ❷

Brachioradialis ❸

Flexor carpi radialis

Flexor carpi ulnaris

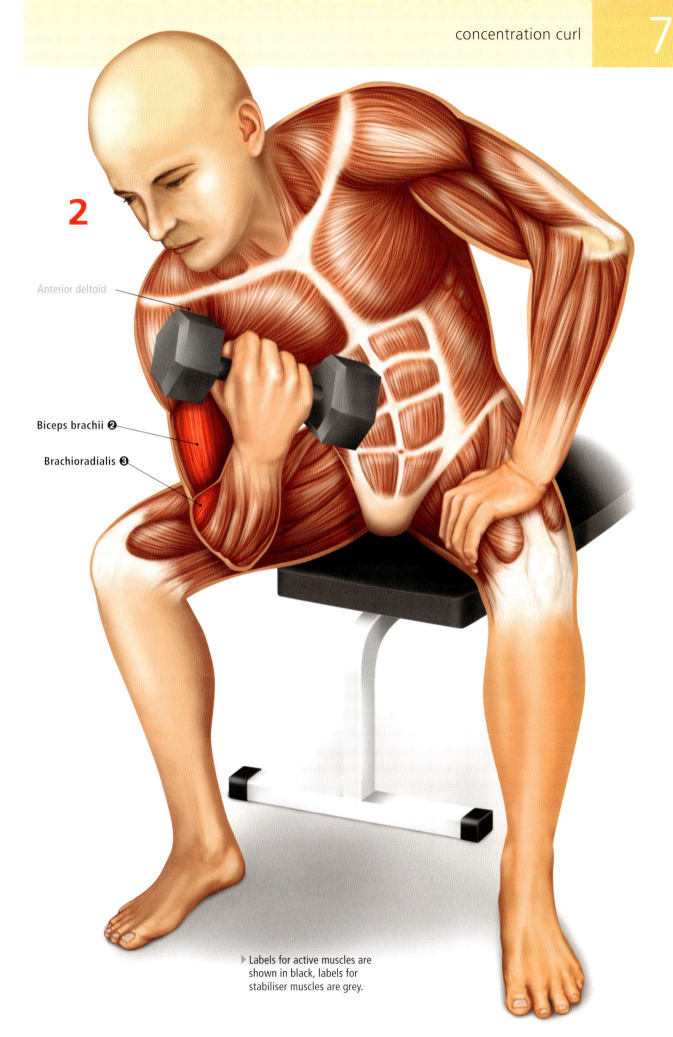

2

Anterior deltoid

Biceps brachii ❷

Brachioradialis ❸

▶ Labels for active muscles are shown in black, labels for stabiliser muscles are grey.

Cable Curl

This biceps exercise is commonly performed at the end of a workout using lighter weights. Brachialis and the long head of biceps brachii are targeted, although as with any exercise in the standing position, the abdominal and lower back muscles are engaged to stabilise the torso, while the rotator cuff muscles must contract to support and stabilise the shoulder. The cable curl is a good exercise for throwing sports, such as baseball, cricket, and water polo, because the biceps muscles are strengthened in a position where they contribute to shoulder stability in the overhead throwing position. This exercise requires two high pulley cables, which makes it best suited to a gym setting.

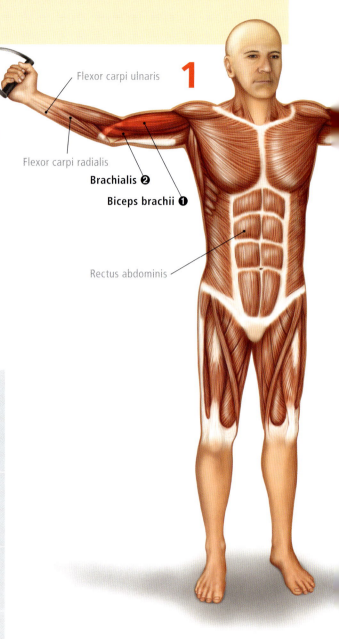

Flexor carpi ulnaris

1

Flexor carpi radialis

Brachialis ❷

Biceps brachii ❶

Rectus abdominis

how to

Using two high pulley cables, grasp each where the cables are far enough apart to stand in the centre with arms outstretched. Palms face towards the ceiling. Curl handles towards the shoulders, keeping upper arms and body still. Slowly return to the starting position until arms are fully extended.

variations

EASY

Grasp the handles of the cables, then take a step backwards so that rather than being straight out to the side, arms are angled forwards at 45 degrees. This takes the stress off the shoulders. Curl handles towards the shoulders, keeping upper arms and body still.

HARD

Use a rope attachment on each pulley for an extra grip challenge. Grasp the rope so that palms face away from the body, then rotate wrists during the lift so that palms face towards the shoulders.

active muscles

❶ Biceps brachii
❷ Brachialis

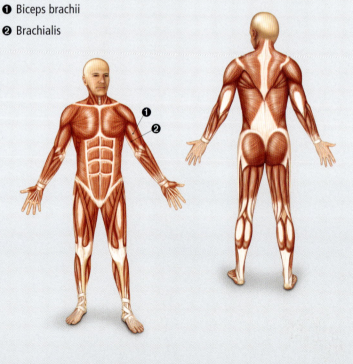

do it right

Ensure that the torso and upper arms stay fixed in place throughout the exercise.

2

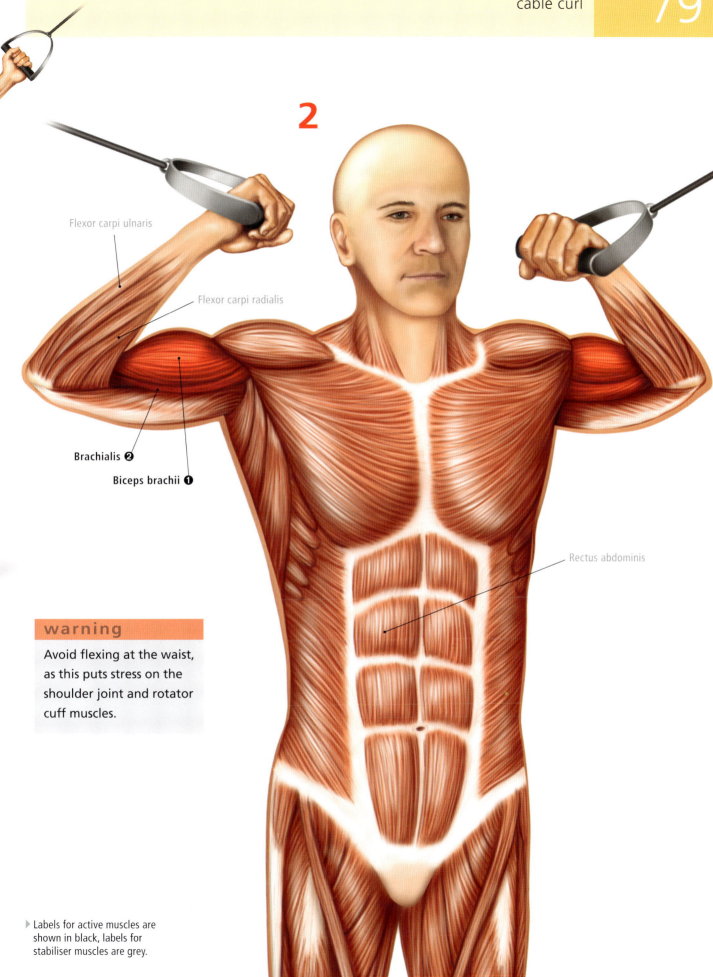

Flexor carpi ulnaris

Flexor carpi radialis

Brachialis ❷

Biceps brachii ❶

Rectus abdominis

warning

Avoid flexing at the waist, as this puts stress on the shoulder joint and rotator cuff muscles.

▷ Labels for active muscles are shown in black, labels for stabiliser muscles are grey.

Triceps Pushdown

This strength-building exercise is a common sight in gyms and fitness centres, performed by people looking to improve the size and strength of their triceps muscles. The versatile pushdown targets the triceps brachii and can be performed with many variations, using different grips and cable attachments. Abdominal muscles get a great secondary workout from this exercise, especially as strength increases and trainers start to lift heavier weights. Suitable for both beginners and the more advanced, the triceps pushdown is best performed in a gym.

how to

Facing a high pulley cable, grasp cable attachment with hands spaced just narrower than shoulder-width apart and palms facing the floor. Extend arms down, pushing the bar towards the floor while keeping elbows by sides and the abdominal muscles tight. Bend at the elbows and return the bar until the forearms are close to the upper arms.

variations

EASY

Stand with one foot forwards and one foot back for extra stability if feeling unsteady or lifting heavy loads. This also helps keep the torso still. It may be more comfortable to use a bar that slopes downwards at 45 degrees on each side, as this takes pressure off the wrists.

HARD

Use a rope attachment on the pulley and grip the rope with thumbs facing the ceiling. Keep hands close together at the top of the movement, rotating the wrists so that palms face the floor at the bottom of the movement. This targets more of the triceps muscle and provides a real grip challenge.

active muscles

❶ Triceps brachii

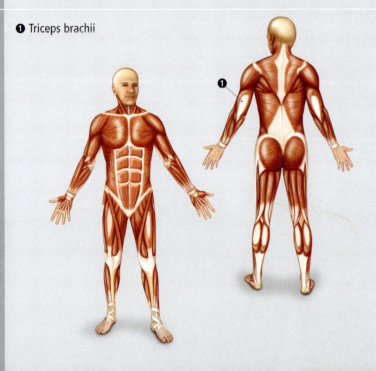

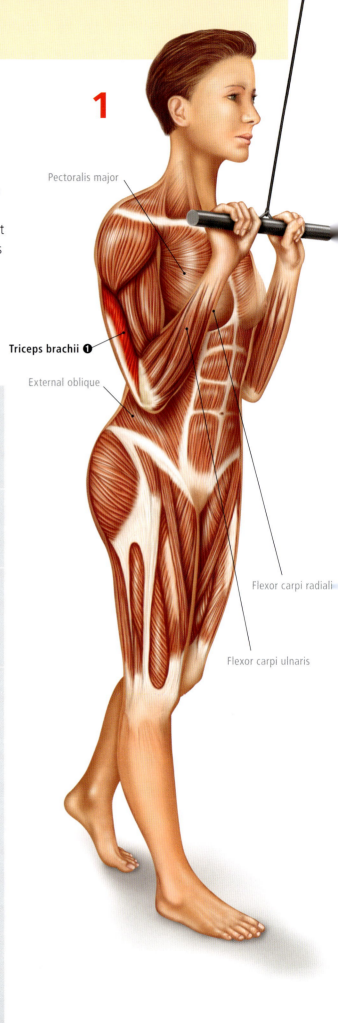

1

Pectoralis major

Triceps brachii ❶

External oblique

Flexor carpi radiali

Flexor carpi ulnaris

2

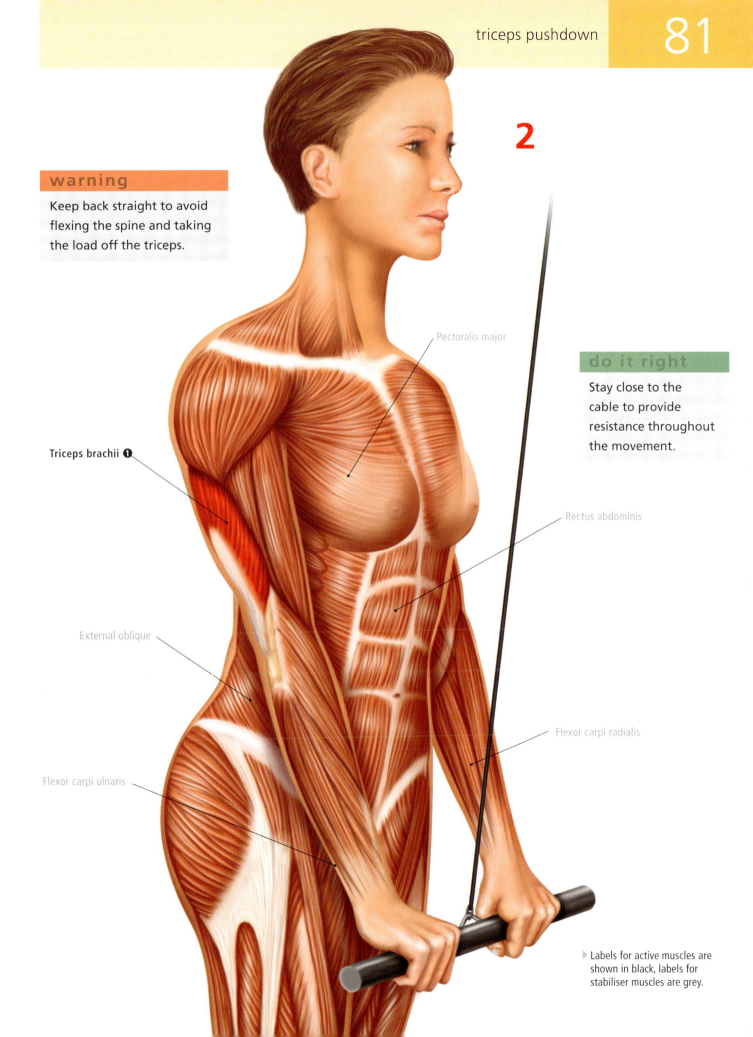

Pectoralis major

Triceps brachii ❶

Rectus abdominis

External oblique

Flexor carpi radialis

Flexor carpi ulnaris

▶ Labels for active muscles are shown in black, labels for stabiliser muscles are grey.

Triceps Extension

This strengthening exercise uses a single dumbbell to effectively target the triceps brachii muscle. The wrist flexor and deltoid help to stabilise the arm, while the triceps works from its most lengthened position, stretching at the bottom of the movement. Anyone training for sports that require strength in overhead positions, such as serving or smashing movements in tennis or volleyball, or in throwing sports, may prefer triceps extensions over triceps pushdowns. This exercise can be performed at home or in a gym, although it requires an attentive spotter to help control the weight, especially when lifting over the head.

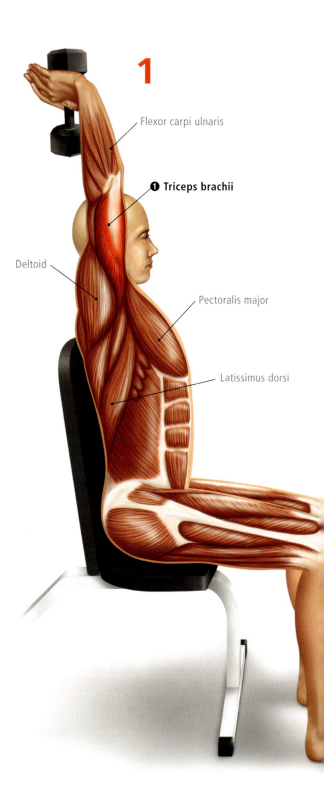

1

Flexor carpi ulnaris

❶ Triceps brachii

Deltoid

Pectoralis major

Latissimus dorsi

how to

Hold a dumbbell with both palms against the plate on one side and thumbs wrapped around the handle. Sitting in a chair or on a bench with back support, start with the dumbbell overhead, arms fully extended. Keep elbows still and lower the weight behind the head by flexing the elbows. Raise dumbbell overhead by extending the elbows.

variations

EASY

Lie back on a bench and grasp one dumbbell in each hand, palms facing each other. Start with arms fully extended towards the ceiling and lower the dumbbells down beside the head, keeping elbows fixed in place. Raise the dumbbells towards the ceiling until arms are again fully extended.

HARD

Perform a one-arm extension by holding the dumbbell with one hand, palm facing forwards. In a sitting position, start with the dumbbell overhead, with arm fully extended. Keep elbow still and lower the weight behind the head by flexing the elbow. Raise the dumbbell overhead by fully extending the elbow.

active muscles

❶ Triceps brachii

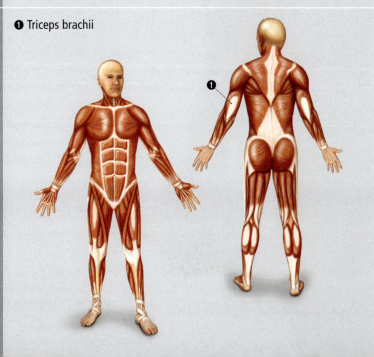

warning

Lower the weight slowly to avoid hitting the back of the neck.

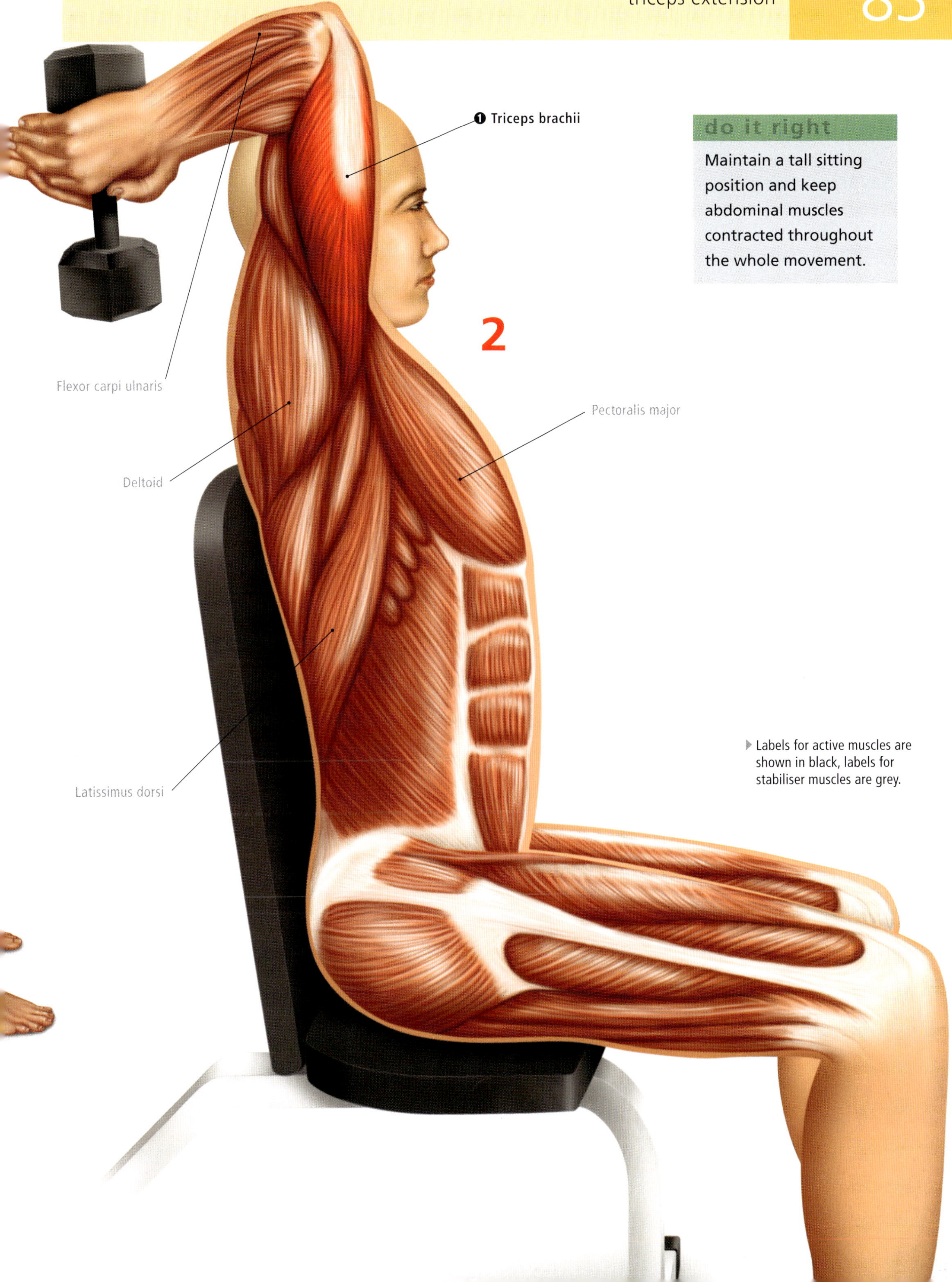

❶ Triceps brachii

Flexor carpi ulnaris

Deltoid

Latissimus dorsi

2

Pectoralis major

do it right
Maintain a tall sitting position and keep abdominal muscles contracted throughout the whole movement.

▶ Labels for active muscles are shown in black, labels for stabiliser muscles are grey.

Triceps Kickback

This simple exercise is performed using a bench and a dumbbell. It targets the triceps brachii at the back of the upper arm. Unlike other triceps exercises, the triceps kickback limits the effective range of motion, making it a suitable exercise for beginners and a good occasional addition to the traditional triceps exercises that more advanced trainers perform. The bench supports the torso, which means there is little or no strain on the lower back. The triceps brachii works hardest when in the shortened position, so trainers will really feel the muscle squeezing at the top of the movement.

warning

Do not rotate the torso to swing the weight up.

how to

Place knee and hand of the supporting arm on a bench so that the torso is horizontal. Position the other foot on the ground, slightly back and to the side for stability. Grasp the dumbbell with palm facing towards the bench, elbow by the side, and arm bent at 90 degrees. Keeping the elbow fixed at the side, extend arm at the elbow until it points straight back. Lower the dumbbell and return to the starting position.

variations

EASY

If experiencing difficulty, lower the elbow so that it is slightly lower than the body. This reduces the range of motion that the triceps has to work through to complete the movement. However, it also reduces the effectiveness of the exercise.

HARD

Grasp the dumbbell with palm facing forwards, elbow by the side and arm bent at 90 degrees. This grip recruits more of the triceps brachii muscle. Keeping the elbow fixed at the side, extend arm at the elbow until it points straight back. Lower the dumbbell and return to the starting position.

active muscles

❶ Triceps brachii

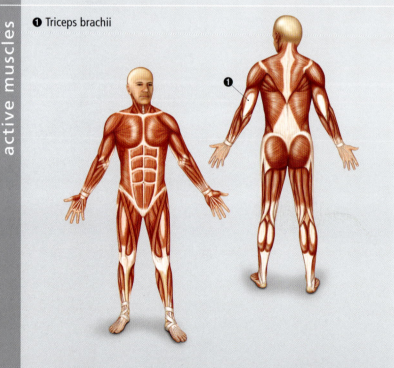

❶ Triceps brachii

Extensor carpi ulnaris

Posterior deltoid

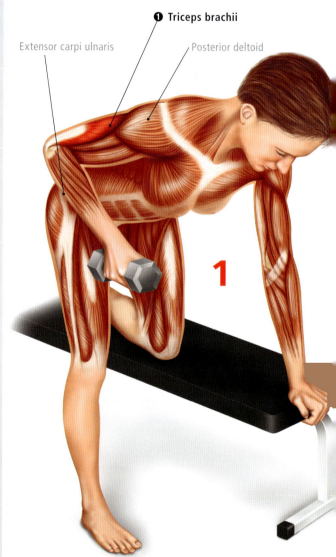

1

2

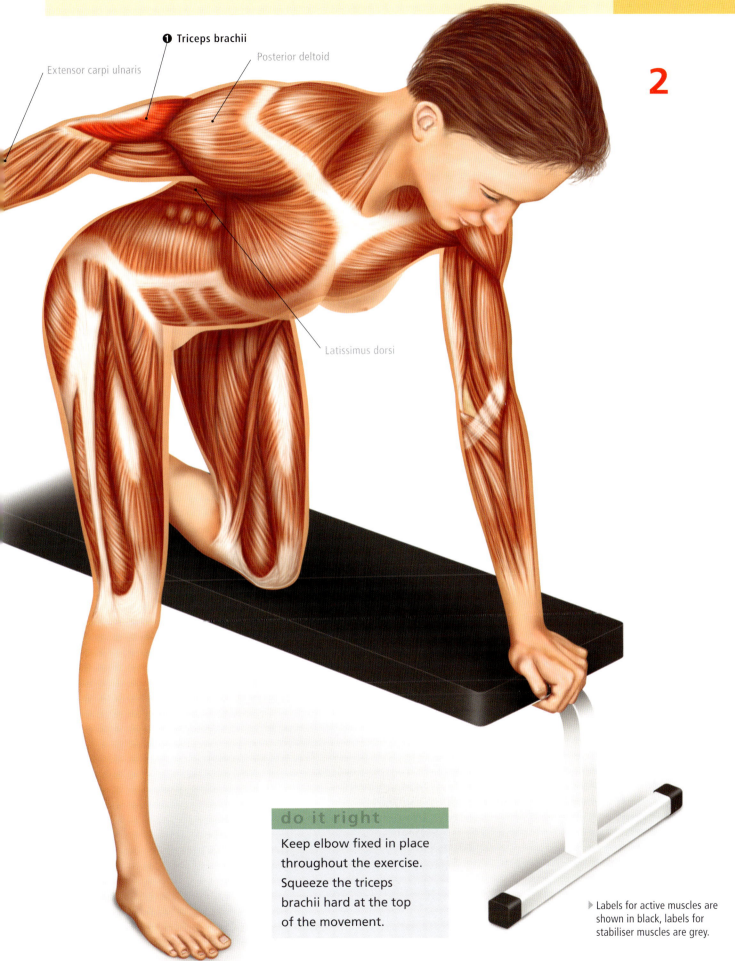

❶ **Triceps brachii**

Posterior deltoid

Extensor carpi ulnaris

Latissimus dorsi

do it right

Keep elbow fixed in place throughout the exercise. Squeeze the triceps brachii hard at the top of the movement.

▶ Labels for active muscles are shown in black, labels for stabiliser muscles are grey.

Shoulder Press

This exercise, and its many variations, is a staple of any complete programme for intermediate and advanced trainers. The shoulder press develops the anterior and lateral deltoids, and provides a good workout for triceps brachii. Lifting dumbbells overhead also requires a great deal of core and shoulder stability, recruiting additional muscles for the task, such as supraspinatus, trapezius, and serratus anterior, the abdominal muscles, and the lower back muscles. This exercise is ideal for sports or occupations that involve pushing or lifting overhead, such as rugby, martial arts, or dancing. It can be performed in a gym or at home but should be done with a training partner to act as a spotter.

warning

Always use a spotter for overhead lifts to prevent injury from dropping weights.

how to

Sit on a bench or chair with back support, feet apart. Grasp each dumbbell, holding them at ear height with palms facing forwards and forearms perpendicular to the ground. Press dumbbells upwards until arms are fully extended overhead, allowing the dumbbells to come together at the top. Slowly lower the weights until they are in the starting position. When lifting heavy weights, use a spotter or a rack to help get the bar into the correct starting position.

variations

EASY

Perform a seated 'military press' on a bench or chair with back support, feet apart. Grasp a barbell with hands slightly wider than shoulder-width apart and palms facing forwards. Press the bar upwards until arms are fully extended overhead. Slowly lower the bar until it is in the starting position.

HARD

Perform an 'Arnold press' by grasping each dumbbell and holding them in front of the shoulders, with palms facing the body and elbows under wrists – like the top position of a chin-up. Press dumbbells upwards and bring elbows out to the sides until arms are fully extended overhead, rotating so palms face forwards. Slowly lower the weights, rotating again until they are in the starting position.

active muscles

❶ Anterior and lateral deltoid

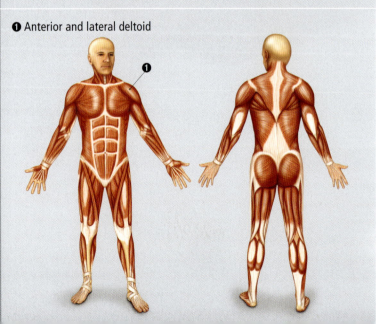

1

Anterior and lateral deltoid ❶

Trapezius

Biceps brachii

Triceps brachii

Serratus anterior

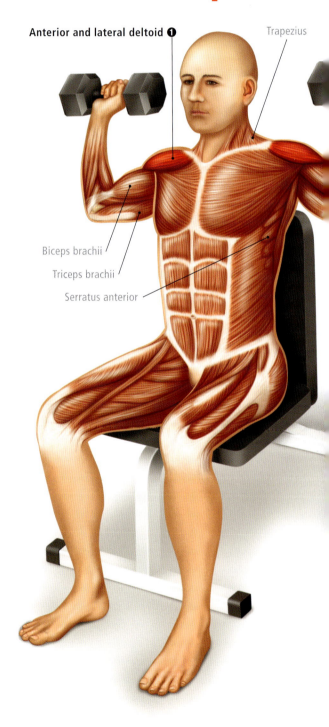

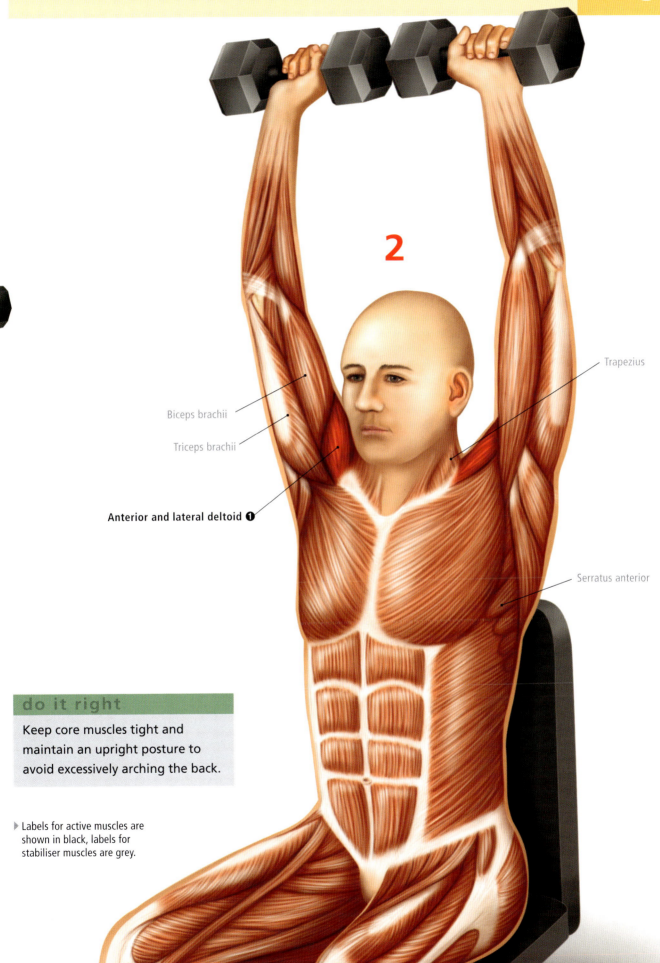

2

Trapezius

Biceps brachii

Triceps brachii

Anterior and lateral deltoid ❶

Serratus anterior

do it right

Keep core muscles tight and maintain an upright posture to avoid excessively arching the back.

▶ Labels for active muscles are shown in black, labels for stabiliser muscles are grey.

Front Raise

This exercise is great for improving strength in the front part of the shoulder. It targets the anterior deltoid muscles, while the rhomboids and trapezius stabilise the shoulder. The abdominals and the lower back muscles also need to work hard to maintain the upright posture, so they get a good secondary workout. The front raise is recommended for sports that require lifting and carrying, as well as combat sports such as martial arts and boxing, which rely on strength and endurance in the anterior deltoid to guard and strike. This exercise may also be used in shoulder rehabilitation, as it moves through a safe and controlled range of motion. Suitable for beginners and advanced trainers, the front raise can be performed at home or in a gym.

how to

Stand with feet shoulder-width apart and knees slightly bent. Grasp the barbell with hands shoulder-width apart, palms facing towards the body, and elbows straight or only slightly bent. Raise the barbell forwards and upwards until the weight is level with the shoulders. Slowly lower the weight down to the starting position.

variations

EASY

Grasp a dumbbell in each hand, palms facing towards the body with elbows straight or only slightly bent. Raise one dumbbell forwards and upwards until the weight is level with the shoulders. Slowly lower the weight down to the starting position, then repeat for the other side.

HARD

Perform the exercise one hand at a time using a low pulley cable. Stand with feet shoulder-width apart and knees slightly bent, facing away from the pulley. Contract the abdominal muscles tightly and raise the cable handle forwards and upwards until the weight is level with the shoulders. This variation puts a greater emphasis on the beginning of the lift.

active muscles

❶ Anterior deltoid
❷ Lateral deltoid
❸ Trapezius

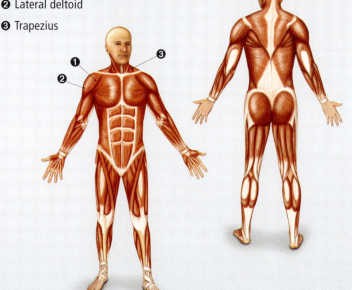

1

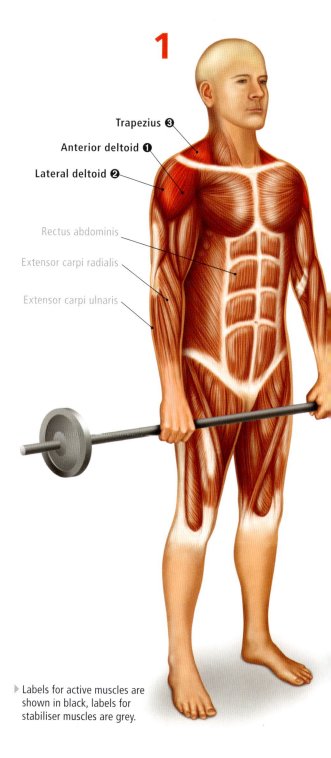

Trapezius ❸
Anterior deltoid ❶
Lateral deltoid ❷

Rectus abdominis
Extensor carpi radialis
Extensor carpi ulnaris

▶ Labels for active muscles are shown in black, labels for stabiliser muscles are grey.

2

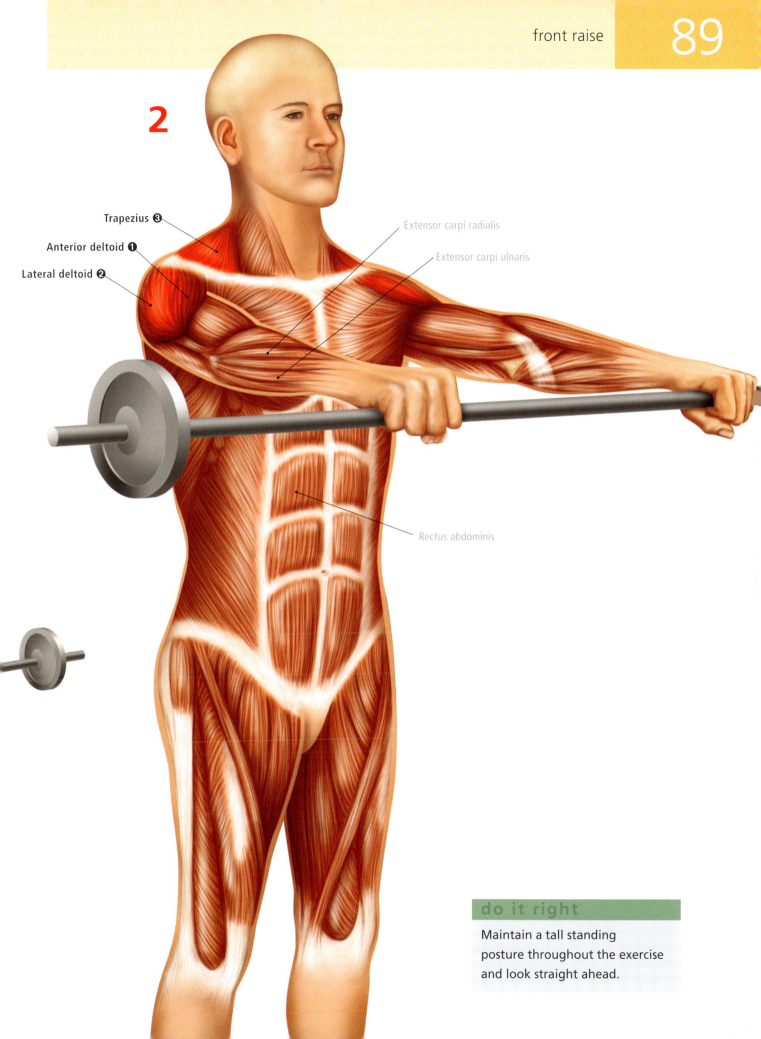

Trapezius ❸
Anterior deltoid ❶
Lateral deltoid ❷

Extensor carpi radialis
Extensor carpi ulnaris

Rectus abdominis

do it right

Maintain a tall standing
posture throughout the exercise
and look straight ahead.

Lateral Raise

This strength-building exercise is designed to specifically target the lateral aspect of the shoulder muscles. The trapezius and wrist extensor muscles also get a great workout, while the rotator cuff muscles are activated to stabilise the shoulder. It is suitable for beginners as well as more advanced trainers. The lateral deltoid is an important muscle for many sporting actions and increasing its size broadens the shoulders, which in effect narrows the appearance of the waist. This makes the lateral raise a perfect programme inclusion for trainers who want to improve their body shapes. As it requires only a set of dumbbells, the lateral raise can be performed both at home and in the gym.

how to

Stand with feet shoulder-width apart, knees slightly bent, and tilt forwards at the hips slightly while maintaining a straight back. Hold a set of dumbbells in front of the thighs, with palms facing towards each other. Raise arms out to the side until elbows are at shoulder height. Make sure elbows remain slightly higher than wrists throughout the movement. Slowly lower the weights to the starting position.

variations

EASY

Train one side at a time if feeling uncomfortable lifting a dumbbell in each hand. Grasp a dumbbell in one hand and hold on to a solid stationary object for support with the other hand. Perform the exercise using the same technique and body position as outlined in the standard version.

HARD

To increase the level of difficulty at the start of the movement, use two low cable pulleys. Stand between two low pulleys; grasp the handle of the right pulley in the left hand, and the left pulley in the right hand. Raise arms out to the side until elbows are shoulder height. Make sure elbows remain slightly higher than wrists throughout the movement.

active muscles

❶ Trapezius

❷ Supraspinatus (under Trapezius)

❸ Lateral deltoid

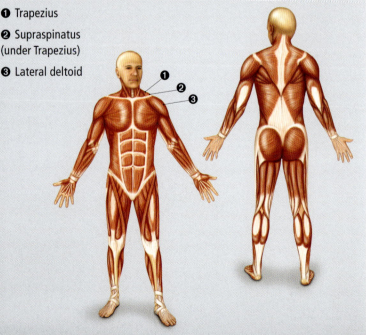

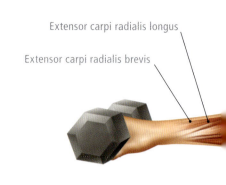

Extensor carpi radialis longus

Extensor carpi radialis brevis

Trapezius ❶

Supraspinatus ❷ (under Trapezius)

Lateral deltoid ❸

Extensor carpi radialis longus

Extensor carpi radialis brevis

1

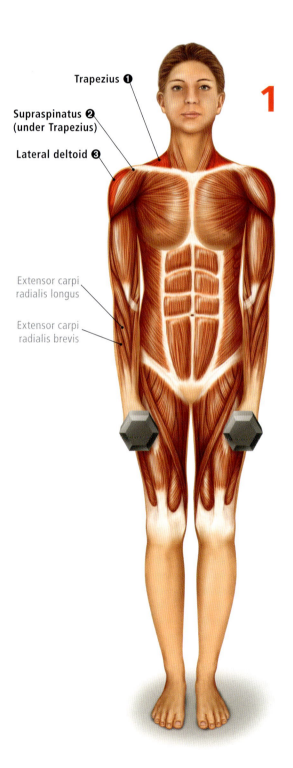

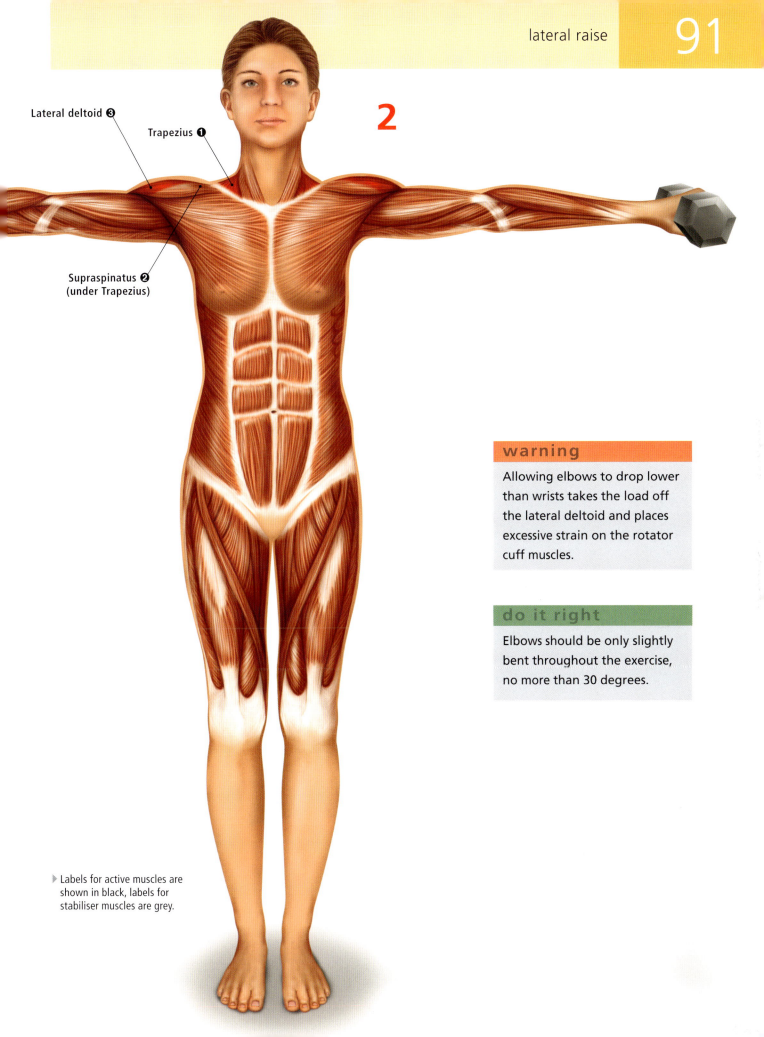

2

Lateral deltoid ❸

Trapezius ❶

Supraspinatus ❷
(under Trapezius)

warning

Allowing elbows to drop lower than wrists takes the load off the lateral deltoid and places excessive strain on the rotator cuff muscles.

do it right

Elbows should be only slightly bent throughout the exercise, no more than 30 degrees.

▷ Labels for active muscles are shown in black, labels for stabiliser muscles are grey.

Upright Row

This is an excellent exercise for improving shoulder strength, and targets the lateral deltoids. The anterior deltoid, trapezius, and biceps muscles are also called on to assist during the lift, while the muscles of the lower back must engage throughout the upright row to provide support to the upper body. This exercise is appropriate for beginners and advanced trainers, and is perfect for sports and occupations that involve lifting, dragging, or pulling movements. It is also a good starting point for trainers who want to learn more advanced, Olympic-style lifts, such as the jerk or snatch.

warning

Do not arch the back or swing back and forth to assist in the lift; this puts strain on the lower back.

how to

Stand with feet shoulder-width apart and knees slightly bent. Grasp the barbell with hands shoulder-width apart, palms facing towards the body, and arms extended. Bend elbows and pull the barbell straight up, keeping the bar close to the body until the elbows are level with the shoulders. Allow the wrists to flex as the bar rises. Slowly lower the bar back to the starting position.

variations

EASY

Perform this exercise one side at a time using a dumbbell. Grasp a dumbbell in one hand and hold on to a solid stationary object for support with the other hand. Use the same technique and body position as the standard upright row, ensuring elbows point out to the side.

HARD

Use two dumbbells instead of a barbell to add extra instability, which requires greater control. Grasp the dumbbells with hands shoulder-width apart, palms facing towards the body, and arms extended. Bend elbows and pull the dumbbells straight up, keeping the weight close to the body until elbows are level with the shoulders. When the dumbbells are raised, wrists should be just below the shoulders with elbows pointed out to the sides.

active muscles

❶ Lateral deltoid
❷ Biceps brachii
❸ Brachioradialis
❹ Trapezius
❺ Supraspinatus (under Trapezius)

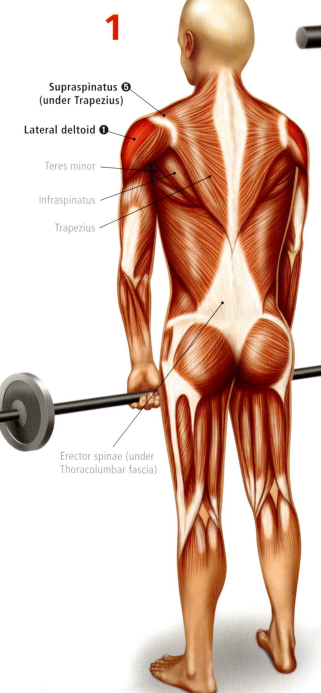

1

Supraspinatus ❺ (under Trapezius)

Lateral deltoid ❶

Teres minor

Infraspinatus

Trapezius

Erector spinae (under Thoracolumbar fascia)

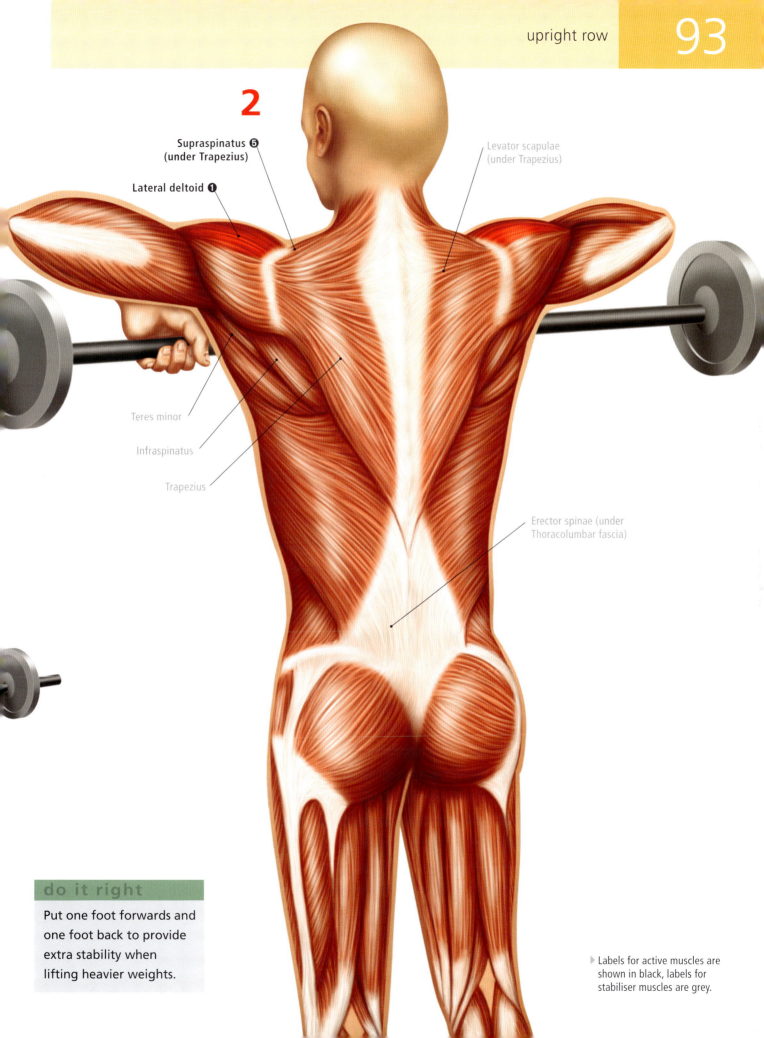

2

Supraspinatus ❺
(under Trapezius)

Lateral deltoid ❶

Levator scapulae
(under Trapezius)

Teres minor

Infraspinatus

Trapezius

Erector spinae (under
Thoracolumbar fascia)

do it right

Put one foot forwards and
one foot back to provide
extra stability when
lifting heavier weights.

▶ Labels for active muscles are
shown in black, labels for
stabiliser muscles are grey.

Shrug

This simple and effective strength-building exercise targets the back of the shoulders and the upper part of the back. The upper fibres of the trapezius muscle do most of the heavy lifting, with levator scapulae assisting. The shrug also gives the forearm muscles a great workout, as the relatively heavy load challenges the grip. This exercise is fantastic for improving strength in movements such as carrying, lifting, and dragging. It suits sports and occupations where these activities are required, and where grip strength is valued, such as rugby, wrestling, and martial arts. A barbell is all that is needed to perform the shrug. A sturdy rack to hold the bar at mid-thigh height will help to avoid injury while getting into the starting position.

1

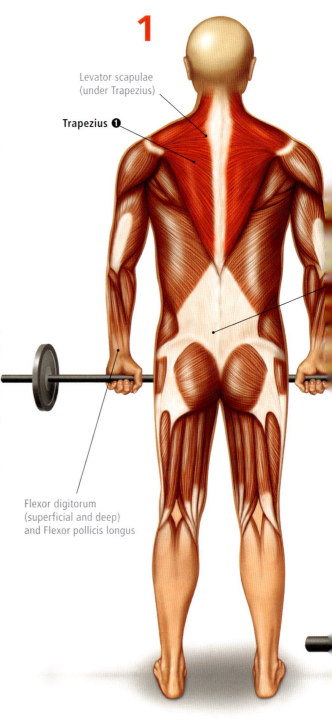

Levator scapulae
(under Trapezius)

Trapezius ❶

Flexor digitorum
(superficial and deep)
and Flexor pollicis longus

how to

Stand with feet shoulder-width apart and knees straight. Grasp the barbell with hands shoulder-width apart, palms facing towards the body, and elbows straight. Allow the shoulders to sink downwards. Pull the barbell straight up by elevating the shoulders towards the ears, keeping the bar close to the body. Slowly lower the bar back to the starting position.

variations

EASY

Use what is called a 'mixed grip', where one palm faces towards the body and one faces away. This allows a stronger grip on the bar and stops the bar from rolling out of the hands. Pull the barbell straight up by elevating the shoulders towards the ears, keeping the bar close to the body. Slowly lower the bar back to the starting position.

HARD

Use one or two low cable pulleys, or two dumbbells, to provide an additional grip challenge and allow for a slightly greater and more natural range of motion. Grasp the dumbbell or pulley handles with hands to the side, below the shoulders, and with palms facing towards the body, elbows straight.

active muscles

❶ Trapezius

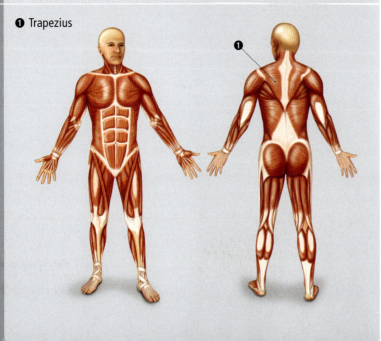

❶

warning

Ensure you use a rack to get the bar into the starting position, and maintain an upright posture during the exercise.

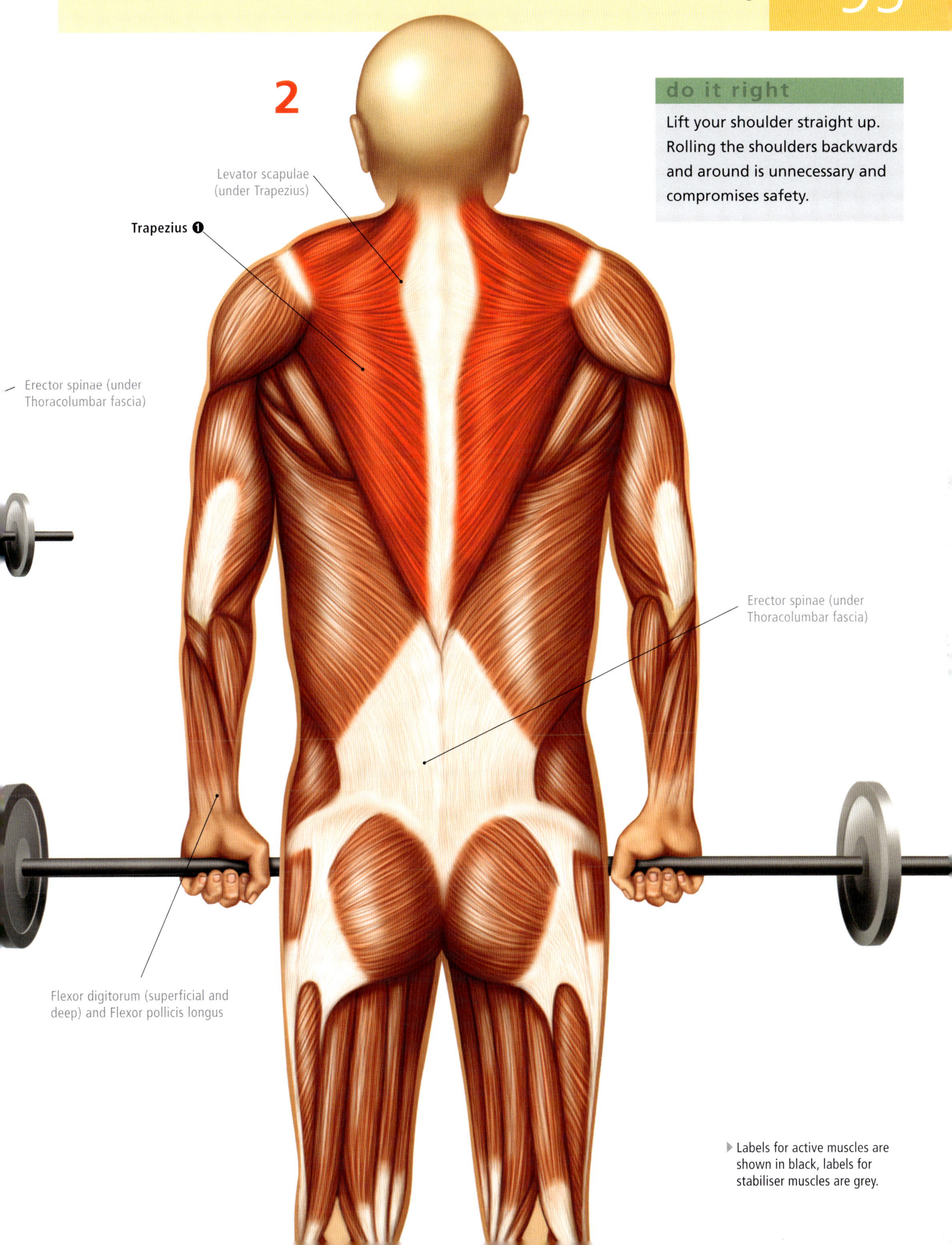

2

Levator scapulae
(under Trapezius)

Trapezius ❶

Erector spinae (under
Thoracolumbar fascia)

do it right

Lift your shoulder straight up.
Rolling the shoulders backwards
and around is unnecessary and
compromises safety.

Erector spinae (under
Thoracolumbar fascia)

Flexor digitorum (superficial and
deep) and Flexor pollicis longus

▶ Labels for active muscles are
shown in black, labels for
stabiliser muscles are grey.

Wrist Curl

This simple exercise focuses on the wrist flexor muscles. The flexor carpi ulnaris and flexor carpi radialis are the main targets, while the muscles that provide grip strength also receive a great workout. Although not necessary as part of a beginner's training programme, the wrist curl is a good addition at the end of a workout for more advanced trainers, to ensure their grip remains strong enough to perform other exercises as they start lifting heavier weights. Sports such as climbing, rugby, and martial arts, as well as all racket and bat sports, benefit from the high level of forearm and grip strength that develops with the wrist curl.

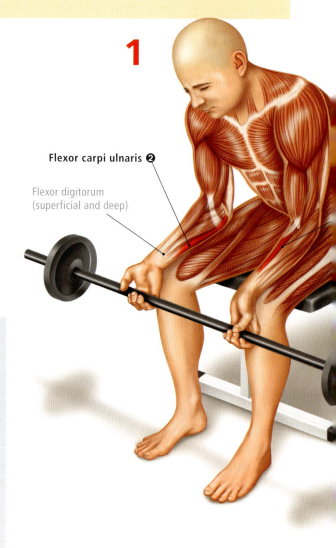

1

Flexor carpi ulnaris ❷

Flexor digitorum
(superficial and deep)

how to

Sit on a bench or chair and grasp a barbell with hands slightly narrower than shoulder-width apart, palms facing the ceiling. Rest forearms on the thighs, with wrists beyond the knees so that they can extend towards the floor without touching the legs. Lower the barbell towards the floor, allowing the bar to roll out of the palms and down into the fingers. Curl the barbell up by gripping the bar into the hand and pointing knuckles towards the ceiling.

variations

EASY

Stand up to perform the wrist curl if the wrist position in extension is uncomfortable. Hold the bar behind the body so that it is just below the gluteal muscles. Grasp the barbell with hands shoulder-width apart and palms facing away from the body. Pull the barbell up by gripping the bar up into the hands and curling the wrists.

HARD

For an extra challenge, perform the wrist curl using one dumbbell in each hand. This requires more control from other muscles in the forearm. Alternatively, perform the exercise with a barbell, palms facing the floor, to work the wrist extensor muscles and give the complete forearm a workout.

active muscles

❶ Flexor carpi radialis
❷ Flexor carpi ulnaris

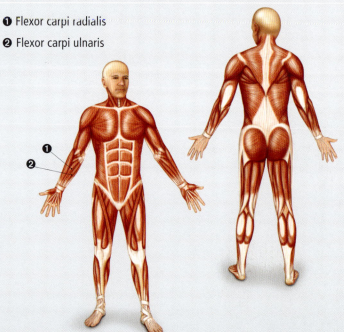

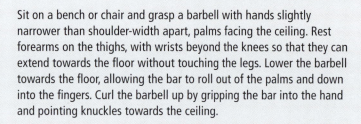

warning

Make sure the floor area is clear so that if the bar rolls out of the fingers it falls safely to the ground, without making contact with anyone's feet.

do it right

Keep wrists and elbows at the same height to maintain resistance on the wrist flexor muscles.

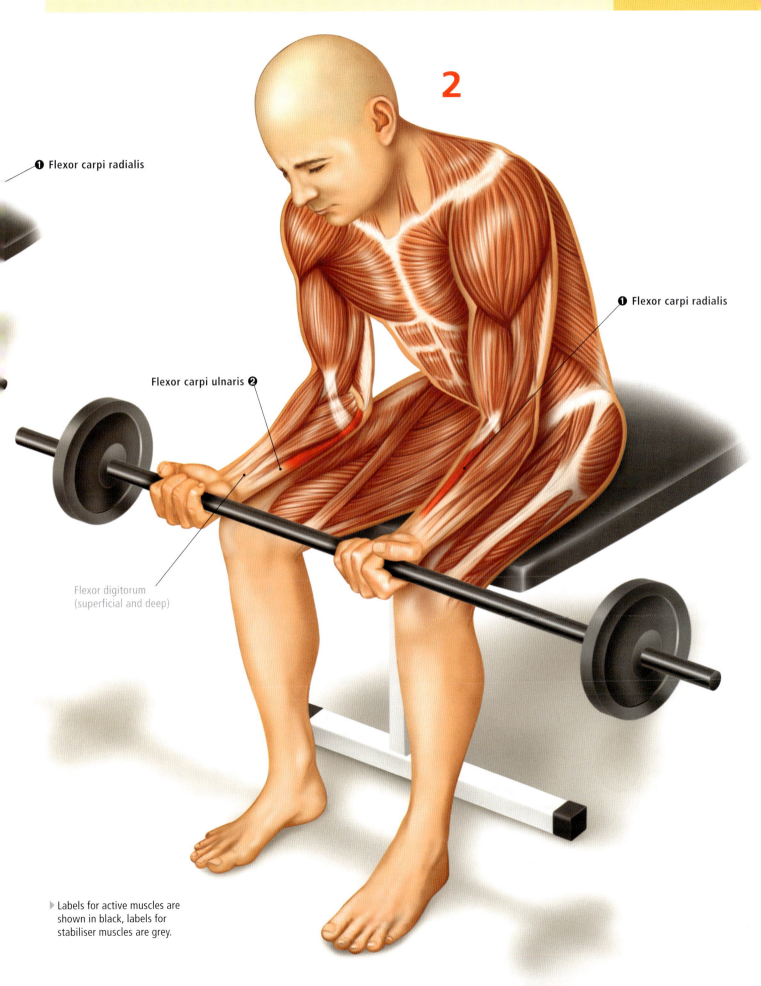

2

❶ Flexor carpi radialis

❶ Flexor carpi radialis

Flexor carpi ulnaris ❷

Flexor digitorum
(superficial and deep)

▷ Labels for active muscles are
shown in black, labels for
stabiliser muscles are grey.

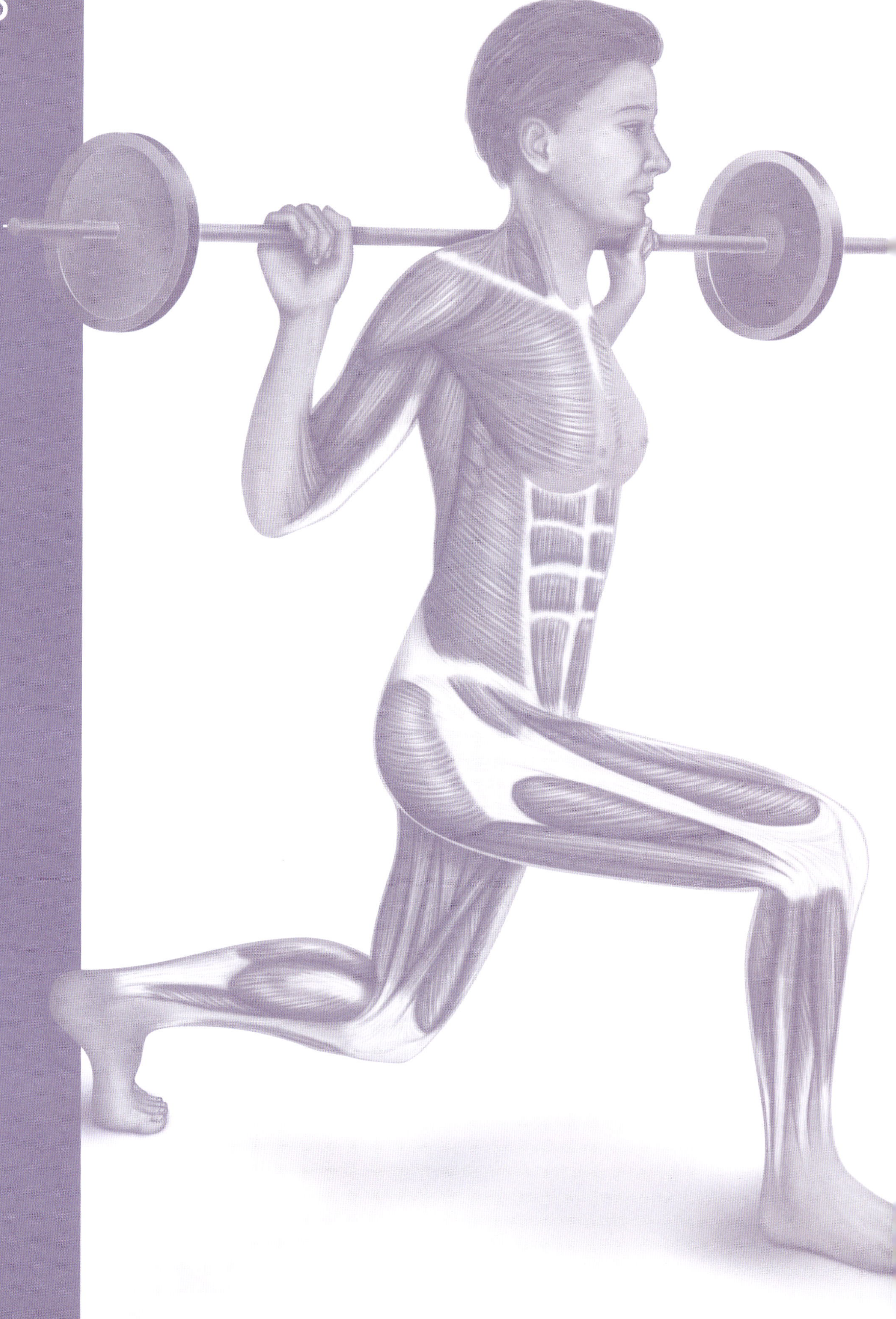

Leg and Buttock Exercises

This section focuses on the muscles required to walk, run, jump, and kick. Weak or injured leg and buttock muscles can cause considerable incapacity. Both beginners and experienced lifters should perform regular resistance training for these regions – to increase muscle strength, tone, power, endurance, and size. The exercises that follow are also suitable for rehabilitation programmes, when allied health professionals prescribe training of specific intensity, volume, and frequency.

Do not neglect exercises that work the backs of the legs, as this can lead to hamstring-quadriceps strength imbalances. Target both sides of the leg to create a balanced workout.

Dumbbell Squat

This effective exercise works the quadriceps, the adductor muscle group, the buttocks, and to a lesser extent the hamstrings and lower back. It is excellent for any sports that require jumping, running, or kicking, and is often used by general fitness enthusiasts who want to firm the lower body. Although suitable for beginners, intermediate, and advanced exercisers, injuries can occur if the dumbbell squat is not performed correctly. Put the emphasis on correct technique. Beginners will feel more stable performing this exercise than the barbell squat, as the dumbbells are held close to the body's centre of gravity. This exercise suits both home and gym environments.

1

how to

Stand with feet shoulder-width or a little wider apart, and hold dumbbells by the sides with palms facing each other. Focus forwards, at about eye level. Bend knees and descend slowly, until the thighs are parallel to the floor. Return to the straight-leg position. Ensure that the back remains flat with a normal lumbar curve throughout the squat, and that heels stay on the floor. If heels start to lift, stop the descent.

variations

EASY

Use a stability ball against a wall, positioned at lower-back level, to provide extra support and balance during the squat. In the descent and ascent phases of the exercise, allow the ball to roll up and down the back, respectively.

HARD

Stand on the edge of a sturdy box or bench and perform a 'single-leg dumbbell squat'. Slowly lower the body as per the standard exercise, but drop one leg below the box and transfer full support of the bodyweight, plus dumbbells if appropriate, to the other leg. Adding external weight will really activate the quadriceps and gluteus maximus. Perform repetitions on each leg.

active muscles

❶ Adductor brevis (under Adductor longus)
❷ Vastus intermedius (under Rectus femoris)
❸ Adductor longus
❹ Adductor magnus
❺ Vastus lateralis
❻ Rectus femoris
❼ Vastus medialis
❽ Gluteus maximus

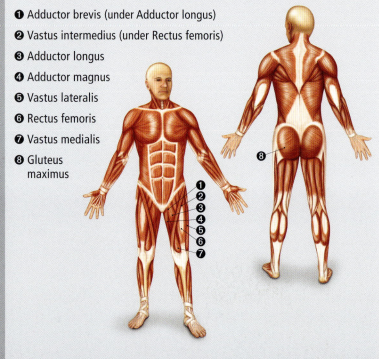

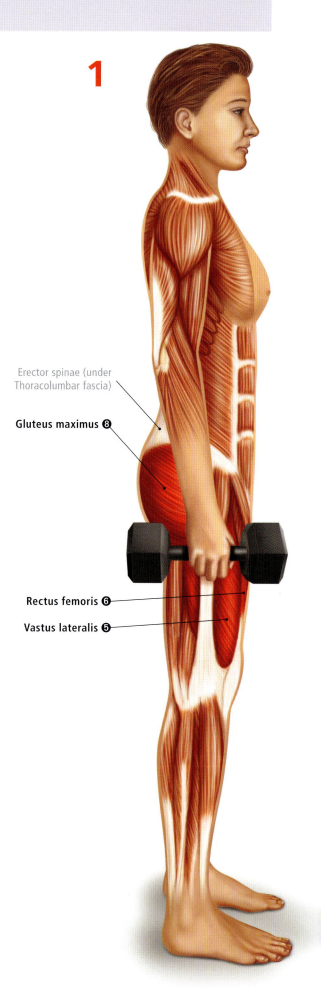

Erector spinae (under Thoracolumbar fascia)

Gluteus maximus ❽

Rectus femoris ❻

Vastus lateralis ❺

2

warning

Do not descend too fast, avoid bouncing out of the bottom of the squat, and do not round the upper back.

Erector spinae (under Thoracolumbar fascia)

Gluteus maximus ❽

❻ Rectus femoris

❺ Vastus lateralis

do it right

Keep heels in contact with the floor at all times. Ensure back is erect by pushing out the chest.

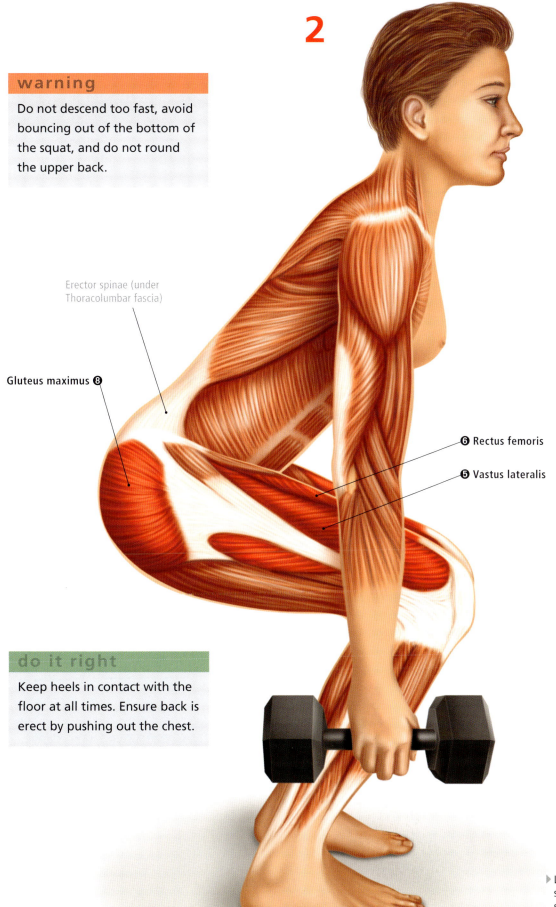

▶ Labels for active muscles are shown in black, labels for stabiliser muscles are grey.

Barbell Squat

This strengthening exercise targets the front of the thighs, the adductors, and the buttocks, as well as the hamstrings and the lower back to a minor extent. In particular, the quadriceps and the gluteus maximus receive a great workout. The barbell squat is ideal for sports that involve jumping, running, or kicking, and is a good lower body firming exercise for trainers. As the bar rests on the upper back during the movement, position it correctly to safeguard against injury. Ensure that proper technique is maintained throughout the exercise before increasing the weight. The barbell squat requires minimal equipment, so it can be performed either at home or in the gym.

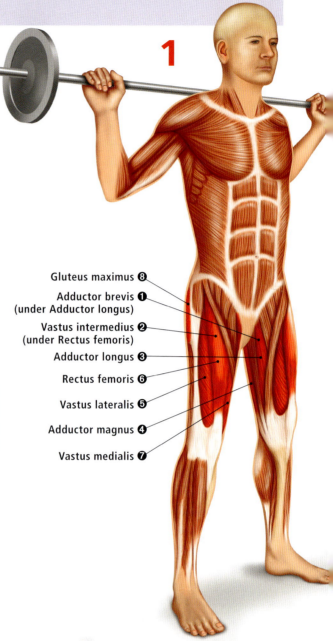

Gluteus maximus ❽
Adductor brevis ❶
(under Adductor longus)
Vastus intermedius ❷
(under Rectus femoris)
Adductor longus ❸
Rectus femoris ❻
Vastus lateralis ❺
Adductor magnus ❹
Vastus medialis ❼

how to

Stand with feet shoulder-width apart, place the barbell across the shoulders, and rest the bar on the upper level of the trapezius muscle. Grip the bar slightly wider than shoulder-width apart. Focus forwards, at about eye level – do not look down. Bend knees and descend slowly, keeping heels on the floor. Stop when thighs are parallel to the floor, then return to the straight-leg position. If heels do start to lift, stop the descent at that point.

variations

EASY

Try a 'quarter squat', which uses the same technique as the standard exercise, but limits the range of motion. Stop the descent when thighs are halfway between being upright and being parallel to the floor, then return to the starting position.

HARD

Try a 'front squat'. Before starting, significantly reduce the amount of weight being lifted. Rest the barbell across the front of the shoulders, with palms facing towards the body and fingertips just gripping the bar. This grip extends the wrist muscles. Follow the standard exercise, but keep the torso in a more upright position to avoid the barbell slipping off the shoulders. Ensure elbows face forwards and upper arms stay parallel to the ground.

active muscles

❶ Adductor brevis
(under Adductor longus)

❷ Vastus intermedius
(under Rectus femoris)

❸ Adductor longus

❹ Adductor magnus

❺ Vastus lateralis

❻ Rectus femoris

❼ Vastus medialis

❽ Gluteus maximus

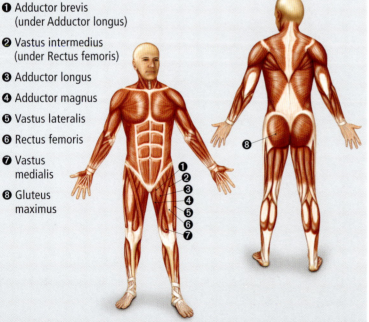

do it right

Keep the back flat, with its normal lumbar curve, throughout the movement. Avoid rounding back forwards during descent or ascent.

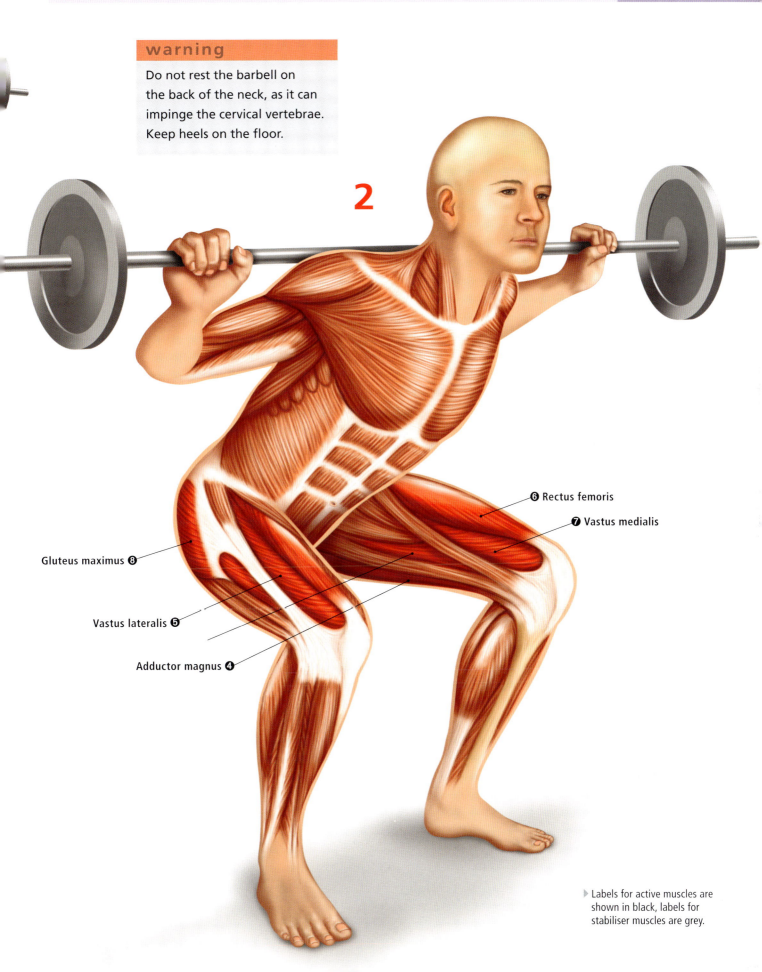

2

❻ Rectus femoris

❼ Vastus medialis

Gluteus maximus ❽

Vastus lateralis ❺

Adductor magnus ❹

▷ Labels for active muscles are shown in black, labels for stabiliser muscles are grey.

Barbell Lunge

This popular exercise works the quadriceps, buttocks, and adductor muscle groups, which include the vastus lateralis, vastus intermedius, vastus medialis, gluteus maximus, adductor brevis, and adductor magnus muscles. It is great for sports that encompass running, jumping, and kicking, as the lunge involves movement at the hip, knee, and ankle joints. Although lighter weights are used in this exercise compared to the various squats, the advantage of the lunge is that it involves stepping out with a single leg – a common movement in many sports. It is suitable for both beginner and advanced exercisers, and can be performed with a barbell, dumbbells, or simply using bodyweight in a home or gym environment.

how to

Stand with feet together and hold barbell across the upper trapezius, with hands slightly wider than shoulder-width apart. Take a large step forwards, placing the front foot on the ground with toes facing forwards. Bend first the front knee, then both knees as the body lowers. Stop when the back leg is close to a 90-degree angle, with heel raised and weight on the ball of the foot. Push up to the starting position using the front leg.

variations

EASY

Hold a dumbbell in each hand by the sides, rather than a barbell across the back if finding it hard to stay balanced. This increases stability throughout the movement because the weights are closer to the body's centre of gravity.

HARD

Perform a 'walking lunge'. Follow the standard exercise, but at the end of each lunge step forwards with the back leg so that it assumes the forwards lunge position. Alternatively, try using dumbbells by the sides, or holding a weight plate or medicine ball above the head with arms extended.

active muscles

❶ Adductor brevis (under Adductor longus)
❷ Vastus intermedius (under Rectus femoris)
❸ Adductor magnus
❹ Vastus lateralis
❺ Vastus medialis
❻ Gluteus maximus

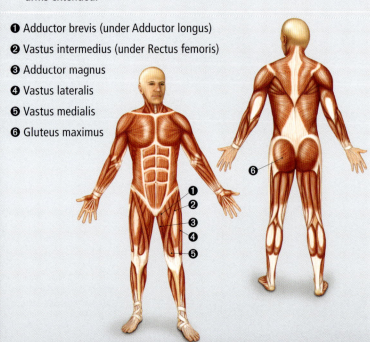

1

Gluteus maximus ❻

Vastus lateralis ❹

Vastus medialis ❺

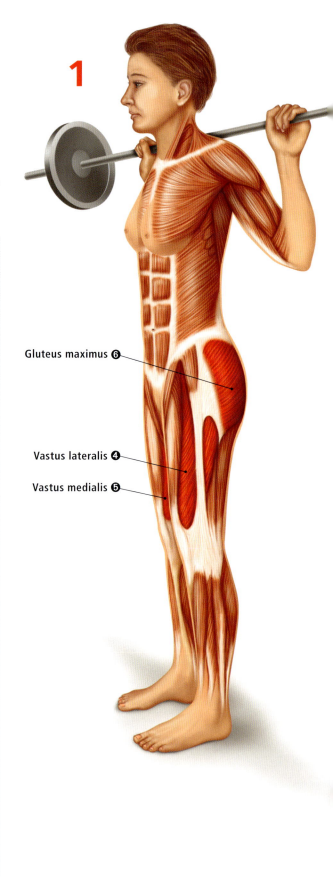

2

do it right

Look straight ahead throughout the movement and maintain an erect torso.

Vastus lateralis ❹

Vastus lateralis ❹

Vastus medialis ❺

❻ Gluteus maximus

❸ Adductor magnus

warning

Do not descend too rapidly. Avoid using heavy weights in the 'walking lunge' variation of this exercise.

▶ Labels for active muscles are shown in black, labels for stabiliser muscles are grey.

Deadlift

This structural exercise works multiple muscles, especially those in the buttocks, legs, and back. When performed with the correct technique, the deadlift is suitable for beginners through to advanced lifters. And because it requires only a barbell and weight plates, it is suitable for both home and gym environments. Those new to resistance training should start with light weights and increase weight slowly, so their bodies can adapt to the increased loading and they can perfect their technique. Advanced lifters can perform the deadlift with heavy weights.

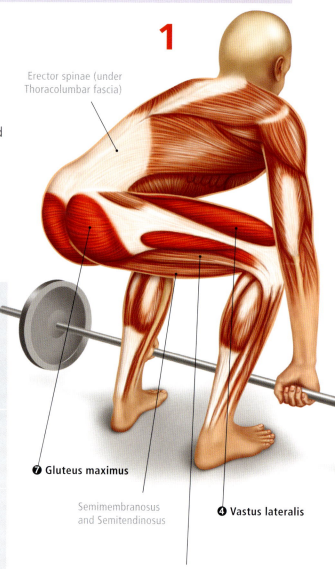

1

Erector spinae (under Thoracolumbar fascia)

❼ **Gluteus maximus**

Semimembranosus and Semitendinosus

❹ **Vastus lateralis**

Biceps femoris

how to

Stand in front of a weighted barbell placed on the ground, with feet shoulder-width apart. Keeping back and arms straight, bend at the knees until able to grasp the bar using an alternating grip, with hands just outside the knees. Look forwards, brace the abdominal and lower back muscles, then straighten knees and hips until fully extended. Shrug shoulders towards the ears at the top of the lift. Lower shoulders, bend knees, and slowly return the barbell to the floor.

variations

EASY

Use a shortened bar, attached to the cable of a pin-loaded weight stack. This makes adjusting the weight easier, and because the weight stack is guided on metal columns, it provides greater stability to the moving bar.

HARD

Increase the challenge by performing a 'sumo deadlift'. Adopt a wide stance with knees bent, feet pointing slightly outwards, and eyes focused directly ahead. Grip the barbell on the inside of the legs, with hands close together in an alternating grip, and arms straight. Keep the back straight, brace abdominal and lower back muscles, then extend the legs. Slowly bend the knees to return the bar to the floor.

active muscles

❶ Adductor brevis (under Adductor longus)
❷ Vastus intermedius (under Rectus femoris)
❸ Adductor magnus
❹ Vastus lateralis
❺ Rectus femoris
❻ Vastus medialis
❼ Gluteus maximus

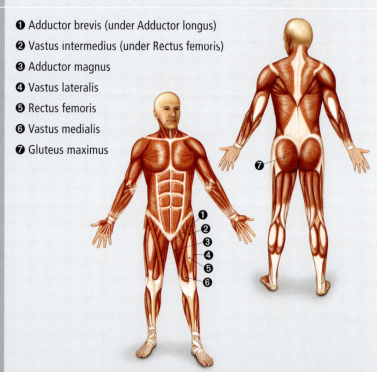

do it right

Maintain a flat back with normal lumbar curve. Keep arms extended throughout the movement and lower weight slowly to the floor.

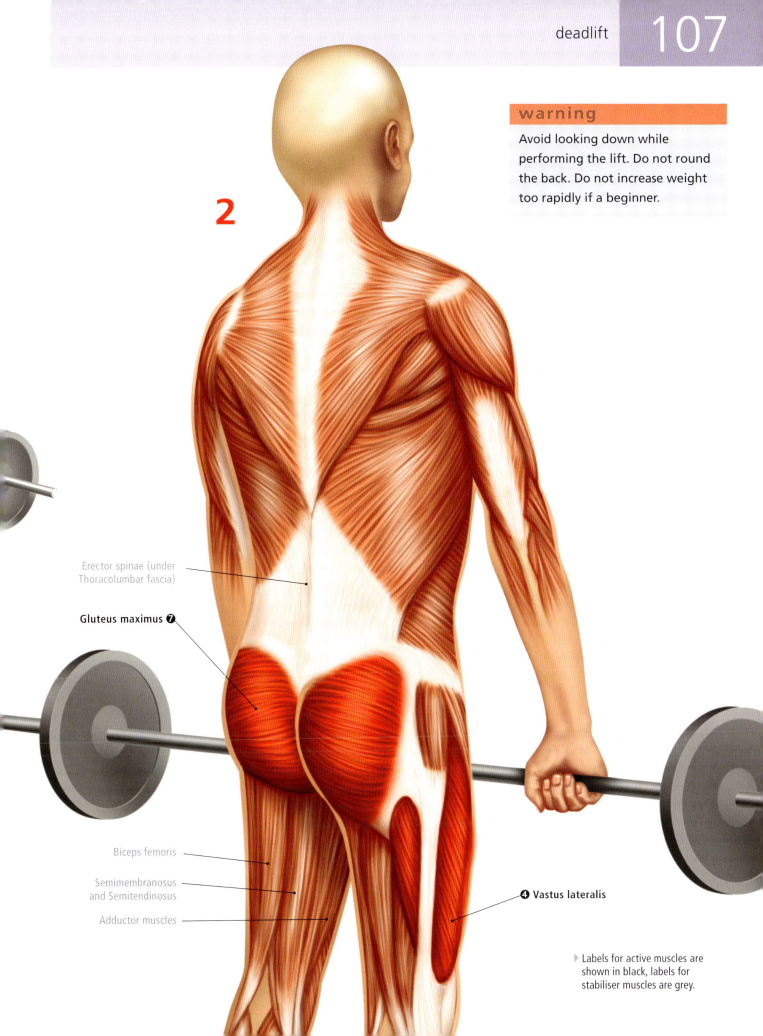

2

Erector spinae (under Thoracolumbar fascia)

Gluteus maximus ❼

Biceps femoris

Semimembranosus and Semitendinosus

Adductor muscles

❹ Vastus lateralis

▶ Labels for active muscles are shown in black, labels for stabiliser muscles are grey.

Romanian Deadlift

This variation of the deadlift is named after the famous Romanian weightlifter, Nicu Vlad, who employed this exercise while training in the United States. It targets the buttocks and hamstring muscle groups, in particular working the gluteus maximus, semitendinosus, semimembranosus, and biceps femoris. Although all levels of lifter can perform the Romanian deadlift, make sure the correct technique is used to avoid injury. In this type of deadlift, there is minimal bending of the knees, so lift less weight in comparison to what is used in other deadlift variations. The Romanian deadlift is suitable for both the home or gym environment, as it requires only a barbell or set of dumbbells to execute.

1

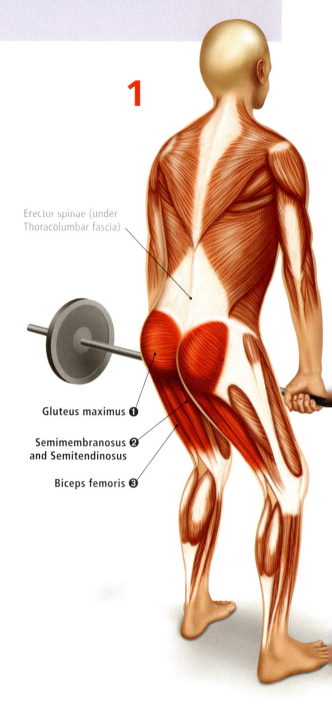

Erector spinae (under Thoracolumbar fascia)

Gluteus maximus ❶

Semimembranosus ❷ and Semitendinosus

Biceps femoris ❸

how to

Stand with feet shoulder-width apart and grasp a barbell with hands just outside the thighs, palms facing towards the body. Bend the knees slightly and retain this knee angle throughout the movement. Keeping the bar close to the body and the back and arms straight, slowly lower the trunk by bending at the hips until the bar reaches a position equal to, or slightly above, the knees. Extend the hips and return to the starting position.

variations

EASY

Use bodyweight only, rather than a barbell, and focus on technique. Once confident that technique is sound, progress to a barbell or try dumbbells instead.

HARD

To increase the challenge, perform a 'single-leg Romanian deadlift'. Hold a dumbbell in each hand at hip level, close to the body, with knees slightly bent. Bend forwards at the hips as per the standard exercise, but allow one leg to extend backwards as the trunk descends. Keep the back straight. Lift the leg until it forms a flat plane with the back. Slowly return to the starting position, then repeat on the other leg.

active muscles

❶ Gluteus maximus

❷ Semimembranosus and Semitendinosus

❸ Biceps femoris

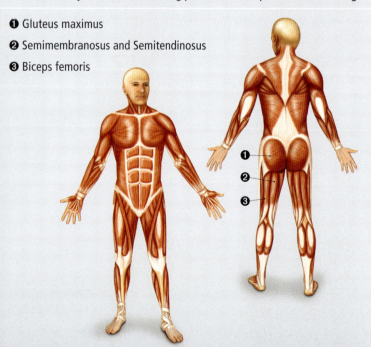

do it right

Keep knees bent at the same angle, and back straight throughout the lift. Lower the trunk in a slow, controlled movement.

2

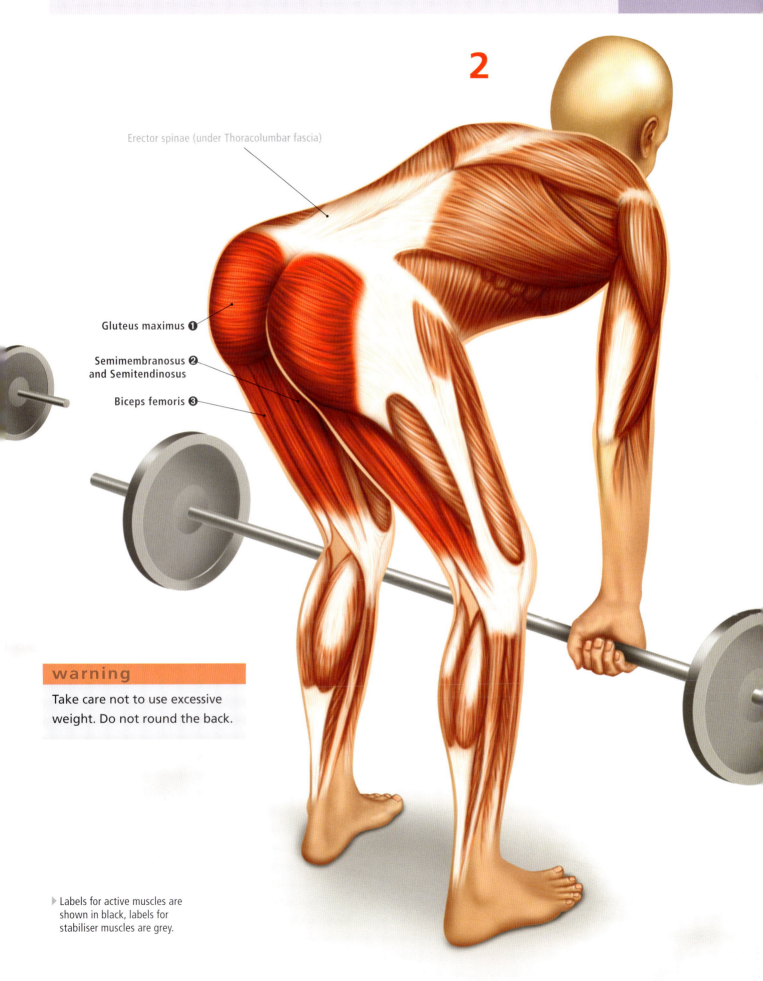

Erector spinae (under Thoracolumbar fascia)

Gluteus maximus ❶

Semimembranosus ❷
and Semitendinosus

Biceps femoris ❸

warning

Take care not to use excessive
weight. Do not round the back.

▶ Labels for active muscles are
shown in black, labels for
stabiliser muscles are grey.

Step-up

This classic exercise focuses on many muscles of the lower body, such as the quadriceps, gluteus maximus, hip flexors (iliopsoas), calves, and to a minor extent the hamstrings. It is therefore a great exercise for sports such as football, hockey, and lacrosse. With a little practice, the step-up is easy to perform. It can be incorporated into most exercise routines and is a popular choice for circuit training programmes. The step-up is suitable for all levels of exercisers. It requires just a barbell or pair of dumbbells, and a sturdy box or step, which makes it easy to perform both at home and in the gym.

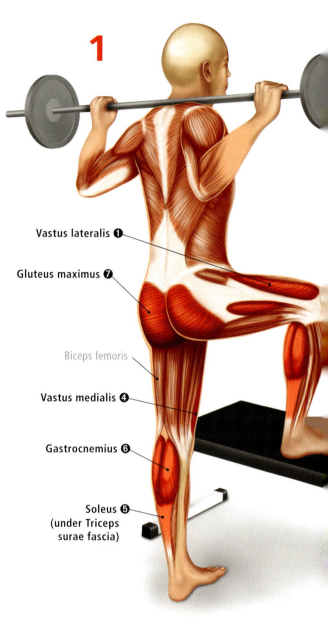

1

Vastus lateralis ❶

Gluteus maximus ❼

Biceps femoris

Vastus medialis ❹

Gastrocnemius ❻

Soleus ❺
(under Triceps
surae fascia)

how to

Stand in front of a sturdy box, ranging from 20–40cm (8–16 inches) in height, with feet close together. Place a weighted barbell on the upper trapezius muscles – avoid contact with the neck. Looking forwards and keeping a straight back, raise one leg off the ground and place foot on the box. Step up, placing the other foot on the box. Step down one leg at a time in a controlled movement.

variations

EASY Hold dumbbells by the sides with palms facing towards the body. This keeps the weight closer to the body's centre of gravity and is therefore less challenging from a balance perspective.

HARD Step up on the box with one foot, but instead of placing the other foot on the box, continue lifting that leg until the thigh is parallel to the ground before placing it down on the box or directly back on the floor. This additional movement requires greater hip flexion and increases the balance challenge.

active muscles

❶ Vastus lateralis

❷ Vastus intermedius
(under Rectus femoris)

❸ Rectus femoris

❹ Vastus medialis

❺ Soleus (under
Triceps surae fascia

❻ Gastrocnemius

❼ Gluteus maximus

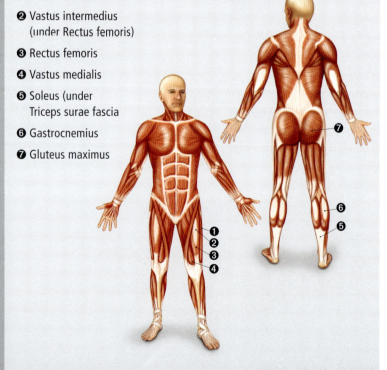

warning

Take care when stepping down from the box. Ensure barbell is off the back of the neck at all times.

2

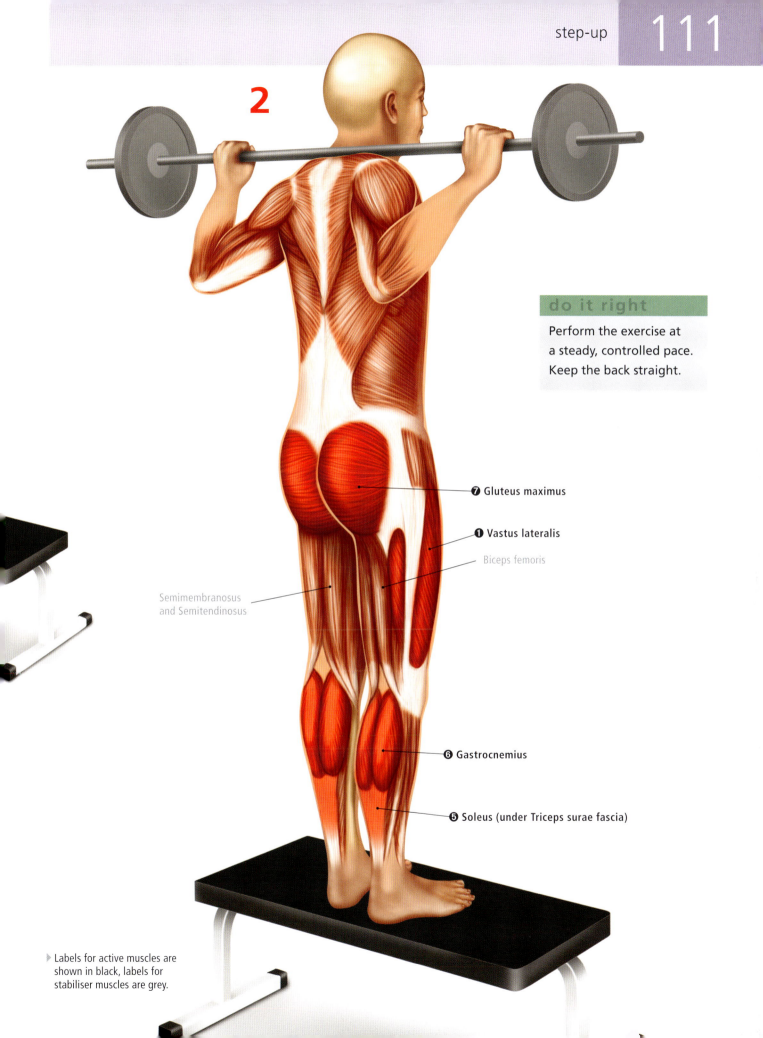

do it right

Perform the exercise at a steady, controlled pace. Keep the back straight.

❼ Gluteus maximus

❶ Vastus lateralis

Biceps femoris

Semimembranosus and Semitendinosus

❻ Gastrocnemius

❺ Soleus (under Triceps surae fascia)

▶ Labels for active muscles are shown in black, labels for stabiliser muscles are grey.

Standing Calf Raise

This leg exercise trains the plantar flexor muscles of the calf, which are used to point the foot in actions such as pushing the accelerator pedal of a car. It is a good exercise choice for beginners through to advanced lifters, and can be performed in a home or gym environment, depending on the variation used and equipment that is available. Experienced lifters often use heavy weights for the standing calf raise. However, beginners should start with light weights and progress only when their technique and strength improve. This exercise benefits sports that involve running or jumping, such as football, hockey, track sprinting, and gymnastics. It is also useful for rehabilitation training for the ankles or calves.

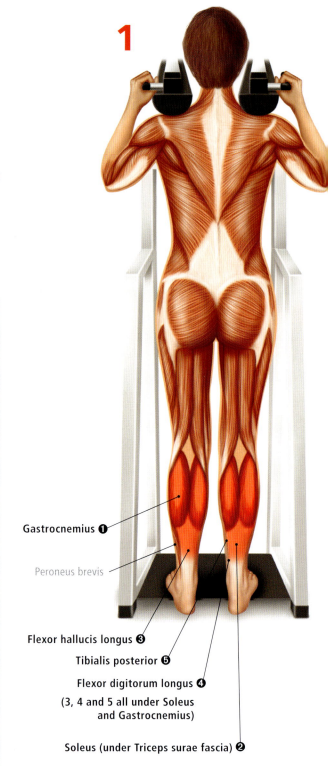

how to

Load the desired weight onto the standing calf-raise machine. Place shoulders on the padded supports and feet on the footplate, with heels slightly protruding from the end of the plate. Keeping a straight back, stand until the legs are extended to assume the starting position. Rise as high as possible on the balls of the feet, then slowly lower the heels to the lowest position possible below the footplate. Do not bend at the knees or hips during the movement.

variations

EASY

Stand with both feet on a step or sturdy box and follow the standard exercise. As strength builds, progress to standing on one leg, then to holding a dumbbell, both of which increase the stress on the calf muscles. This is a good option for home training, when a standing calf-raise machine may not be available.

HARD

Ramp up the workout with eccentric overload training, one leg at a time. Perform the upwards movement with both calves, lifting the weight against gravity, but slowly resist the downwards movement with only one calf, relaxing the opposite calf. Repeat until the desired number of downwards movements has been completed with each calf.

active muscles

❶ Gastrocnemius

❷ Soleus (under Triceps surae fascia)

❸ Flexor hallucis longus

❹ Flexor digitorum longus

❺ Tibialis posterior

❻ Plantaris

(3, 4 and 5 all under Soleus and Gastrocnemius)

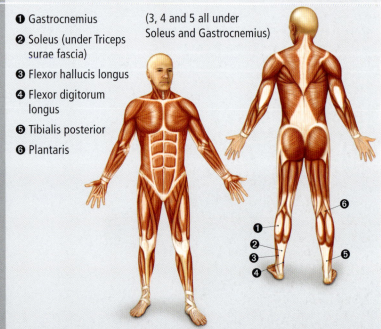

Gastrocnemius ❶

Peroneus brevis

Flexor hallucis longus ❸

Tibialis posterior ❺

Flexor digitorum longus ❹

(3, 4 and 5 all under Soleus and Gastrocnemius)

Soleus (under Triceps surae fascia) ❷

warning

Do not bounce at the bottom of the movement. Stay slow and controlled as heels descend.

2

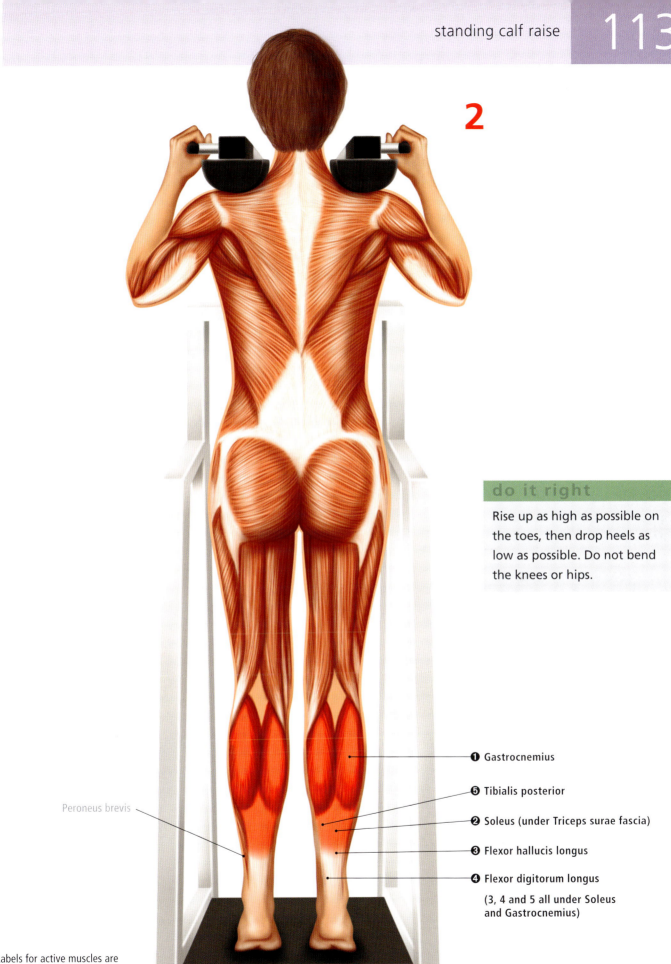

Rise up as high as possible on the toes, then drop heels as low as possible. Do not bend the knees or hips.

❶ Gastrocnemius

❺ Tibialis posterior

❷ Soleus (under Triceps surae fascia)

❸ Flexor hallucis longus

❹ Flexor digitorum longus

(3, 4 and 5 all under Soleus and Gastrocnemius)

Peroneus brevis

▷ Labels for active muscles are shown in black, labels for stabiliser muscles are grey.

Seated Calf Raise

This calf exercise is another good choice for training the plantar flexor muscles. The bent-knee position that it requires minimises the involvement of the large gastrocnemius muscle and places more emphasis on the other plantar flexor muscles of the calf. Use it in training for sports such as track sprinting, football, gymnastics, and hockey. This exercise can also be successfully included in rehabilitation programmes for the calves and ankles. The seated calf raise is appropriate for all levels of trainer and it is possible to perform in a home or gym environment. A specialised machine is generally used for the standard exercise; however, other variations that do not require a machine are just as effective.

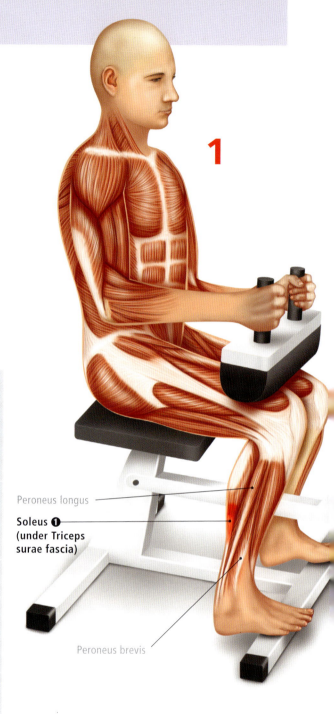

Peroneus longus

**Soleus ❶
(under Triceps
surae fascia)**

Peroneus brevis

how to

Load the desired weight onto the seated calf-raise machine. Sit down and place lower thighs and knees under the padded support, and feet on the footplate, with heels protruding from the back of the plate. Lift the padded support by rising onto the balls of the feet. Slowly lower the weight until heels drop as far below the footplate as possible. Repeat sets as desired.

variations

EASY

Sit on a chair or bench and place feet on a low box or step. Place a folded towel across the lower thighs and knees, then rest a loaded barbell on the towel. Lift the heels until balancing on the balls of the feet. Slowly lower the heels as far as possible below the box.

HARD

Increase the challenge by focusing on one calf at a time. Perform the upwards movement as per the standard exercise, with both calves lifting the weight against gravity. Slowly resist the downwards movement with only one calf, relaxing the opposite calf. This applies eccentric overload to the muscles in the lowering phase of the exercise. Repeat with the alternate leg.

active muscles

❶ Soleus (under Triceps surae fascia)

❷ Flexor hallucis longus

❸ Flexor digitorum longus

❹ Tibialis posterior

❺ Plantaris

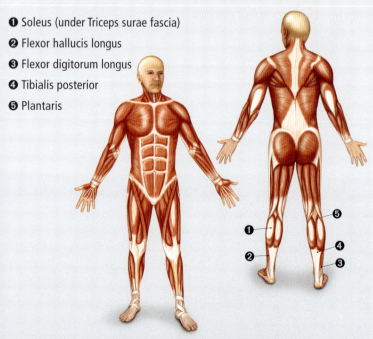

do it right

Rise as high as possible on the balls of the feet during the upwards movement, and drop heels as low as possible during the downwards movement.

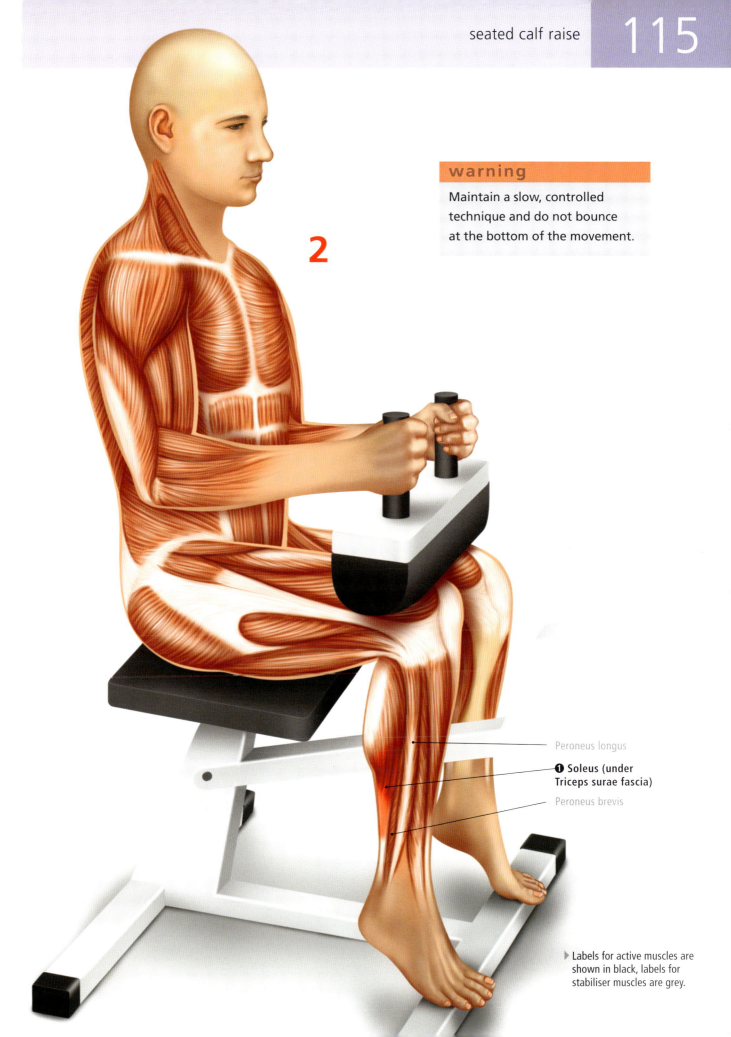

warning

Maintain a slow, controlled technique and do not bounce at the bottom of the movement.

2

Peroneus longus

❶ **Soleus (under Triceps surae fascia)**

Peroneus brevis

▸ **Labels** for active muscles are shown in black, labels for stabiliser muscles are grey.

Leg Extension

This popular exercise targets the quadriceps muscle group. It is useful for any sports that involve running, kicking, jumping, or skipping movements. The leg extension is suitable for beginners through to advanced exercisers, and can be executed in a home or gym environment, depending on which variation is used. As it is generally performed seated on a specialised machine, this exercise does not require the same level of balance as squat and lunge exercises. Although this can be an advantage in some circumstances, such as early rehabilitation scenarios where balance may be problematic, it does make the leg extension less useful for trainers who are looking for balance and stability challenges in their routines.

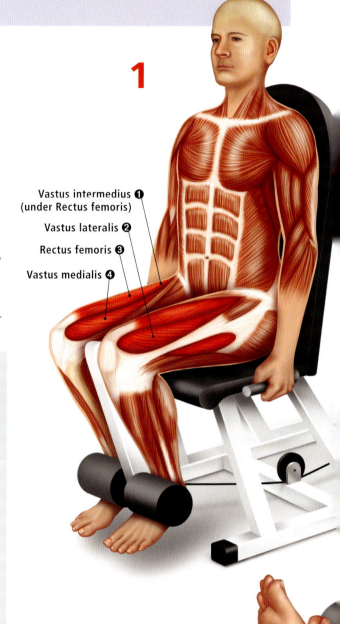

1

Vastus intermedius ❶
(under Rectus femoris)

Vastus lateralis ❷

Rectus femoris ❸

Vastus medialis ❹

how to

Sit on a leg-extension machine and adjust the back rest so the back of the knees rest against the edge of the padded seat and knee joints align with the rotation axis of the lever arm. Adjust the leg pad to just above the front of the ankles. Pushing against the leg pad, lift legs until fully extended. Keep the torso upright and feet pointing forwards. Slowly return to the starting position, then repeat.

variations

EASY

Use ankle weights and perform the exercise sitting on a chair or sturdy table. Position the weight cuff directly above the ankle of each leg. If using a chair, ensure that it is tall enough to avoid feet touching the ground on the return phase of the movement. Keep torso straight throughout the exercise.

HARD

Go for the extra challenge of eccentric overload when lowering weight, one leg at a time. Perform the standard exercise, using both legs to raise the weight until legs are fully extended. However, on the return phase of the movement, resist the pull of the weight with only one leg. Complete repetitions on each leg.

active muscles

❶ Vastus intermedius (under Rectus femoris)

❷ Vastus lateralis

❸ Rectus femoris

❹ Vastus medialis

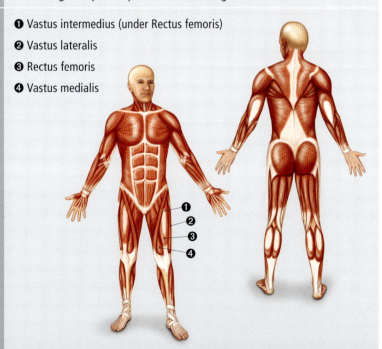

warning

Do not descend too rapidly or swing the trunk.

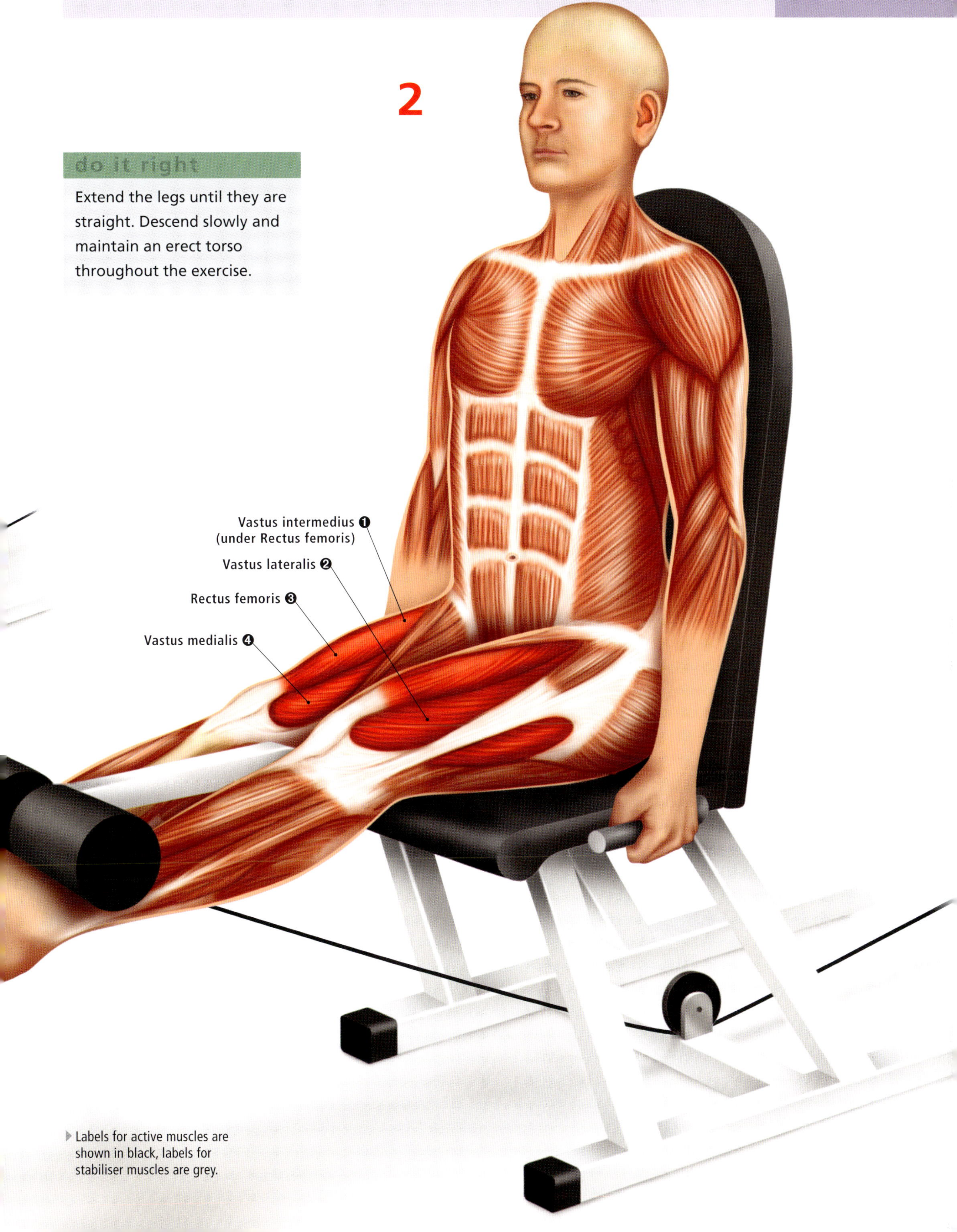

2

do it right

Extend the legs until they are straight. Descend slowly and maintain an erect torso throughout the exercise.

Vastus intermedius ❶
(under Rectus femoris)

Vastus lateralis ❷

Rectus femoris ❸

Vastus medialis ❹

▶ Labels for active muscles are shown in black, labels for stabiliser muscles are grey.

Seated Leg Curl

This seated exercise is another popular choice for homing in on the semitendinosus, semimembranosus, and biceps femoris muscles of the hamstrings. Like the lying leg curl, it provides great training for sports that involve running and kicking movements, such as track, field, and football. The seated leg curl is suitable for beginners and advanced exercisers; however, because it requires equipment to execute, it is not usually performed at home. Many trainers prefer this exercise to the lying leg curl because they find the seated position more comfortable.

how to

Sit in a seated leg-curl machine and adjust the back support so that knees rest against the edge of the padded seat, with knee joints aligned with the rotation axis of the lever arm. Level leg pads with the back of the ankle. Hold the handgrips and pull against the leg pads, curling legs as far as the machine will allow towards the buttocks. Slowly return to the starting position and repeat. Keep torso upright and feet pointing forwards throughout the exercise.

variations

EASY

Before beginning, reduce the amount of weight to the lightest plate possible, then follow the standard exercise.

HARD

Follow the standard exercise, curling both legs towards the buttocks. However, on the return phase, resist the pull of the weight with only one leg. This places greater stress on the hamstring muscles during the lengthening phase, as this motion (eccentric phase) is capable of greater force than the shortening phase (concentric phase) when the legs move towards the buttocks. Alternate legs until the desired number of repetitions have been completed.

active muscles

❶ Semitendinosus
❷ Semimembranosus
❸ Biceps femoris

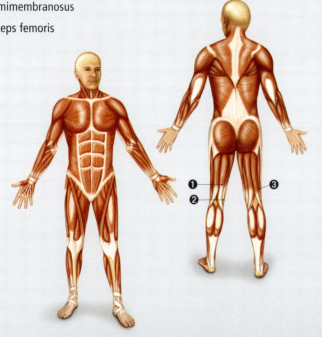

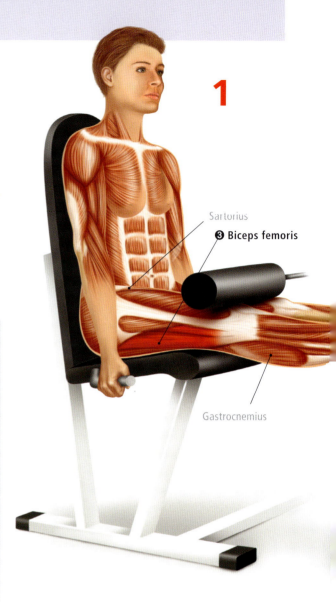

Sartorius

❸ Biceps femoris

Gastrocnemius

1

do it right

Curl the legs towards the buttocks as far as the leg pad will allow. Return slowly to the starting position and maintain an erect torso.

2

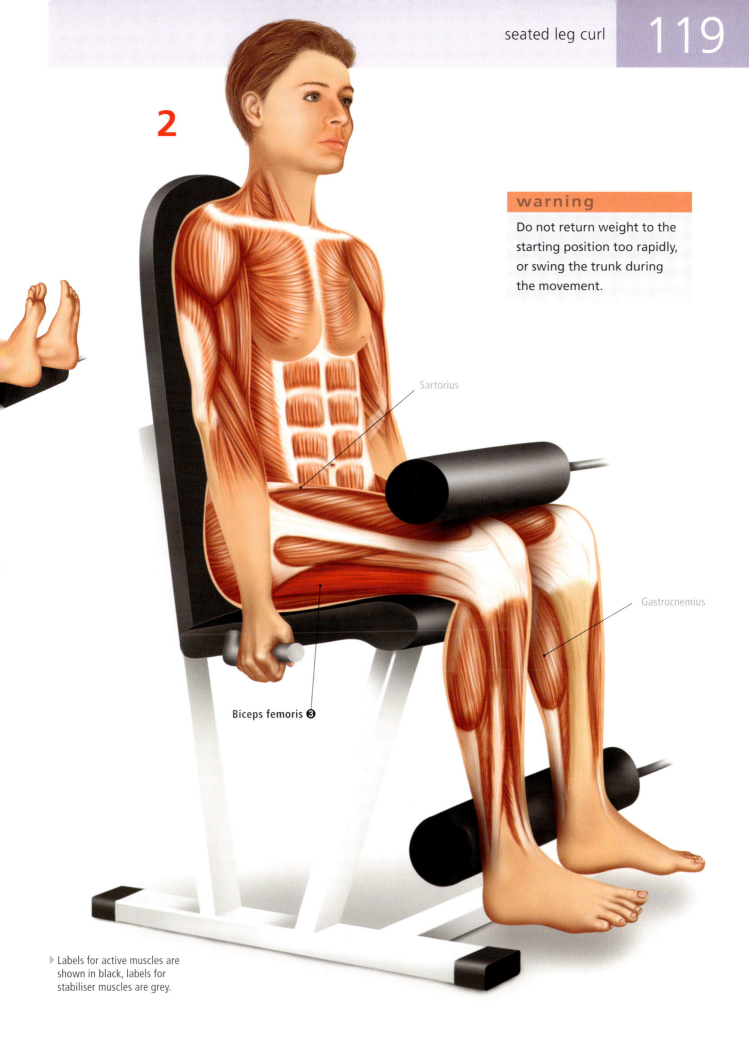

warning

Do not return weight to the starting position too rapidly, or swing the trunk during the movement.

Sartorius

Gastrocnemius

Biceps femoris ❸

▷ Labels for active muscles are shown in black, labels for stabiliser muscles are grey.

Lying Leg Curl

This popular exercise really works the semitendinosus, semimembranosus, and biceps femoris in the hamstring muscle group. Targeting this area assists running and kicking movements, which make the lying leg curl a good inclusion for most sports-specific programmes. It is suitable for beginners and advanced exercisers. The standard version of this exercise requires a specialised machine, which restricts its execution to commercial gyms or well-equipped homes. However, the easy variation uses ankle weights, making it possible to perform the lying leg curl in a simple home environment. This variation is also a great inclusion for rehabilitation training programmes that focus on the hamstrings.

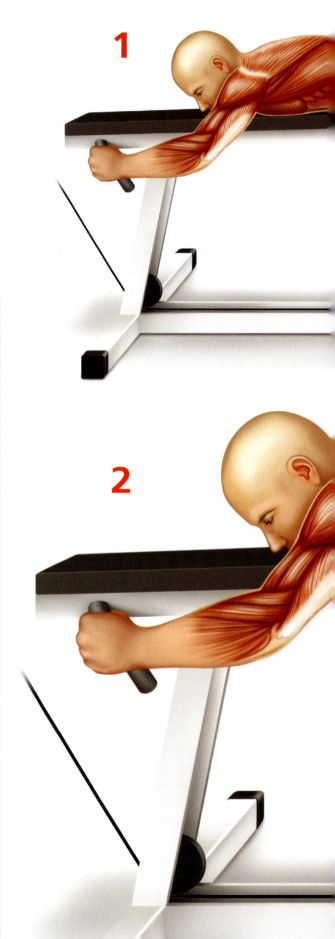

1

2

how to	Lie face down on the machine with knees slightly protruding from the edge of the padded surface, so knee joints line up with the rotation axis of the lever arm. Adjust pad behind legs to just above ankle level. Curl legs up as far as possible towards the buttocks. Take two or three seconds to slowly return to the starting position with legs extended. Keep upper body in contact with pads at all times.
variations **EASY**	Use ankle weights and lie face down on a sturdy bench or table, with the knees just protruding from the edge. Perform movement as per the standard exercise. This variation is not restricted to a gym setting, but it does limit how much weight can be lifted.
HARD	Target the eccentric phase of the movement, one leg at a time. Perform the exercise initially as per the standard movement, curling both legs towards the buttocks. Then, on the return phase, resist the pull of the weight with only one leg. Complete the desired number of eccentric overload repetitions on each leg.
active muscles	❶ Biceps femoris ❷ Semimembranosus and Semitendinosus

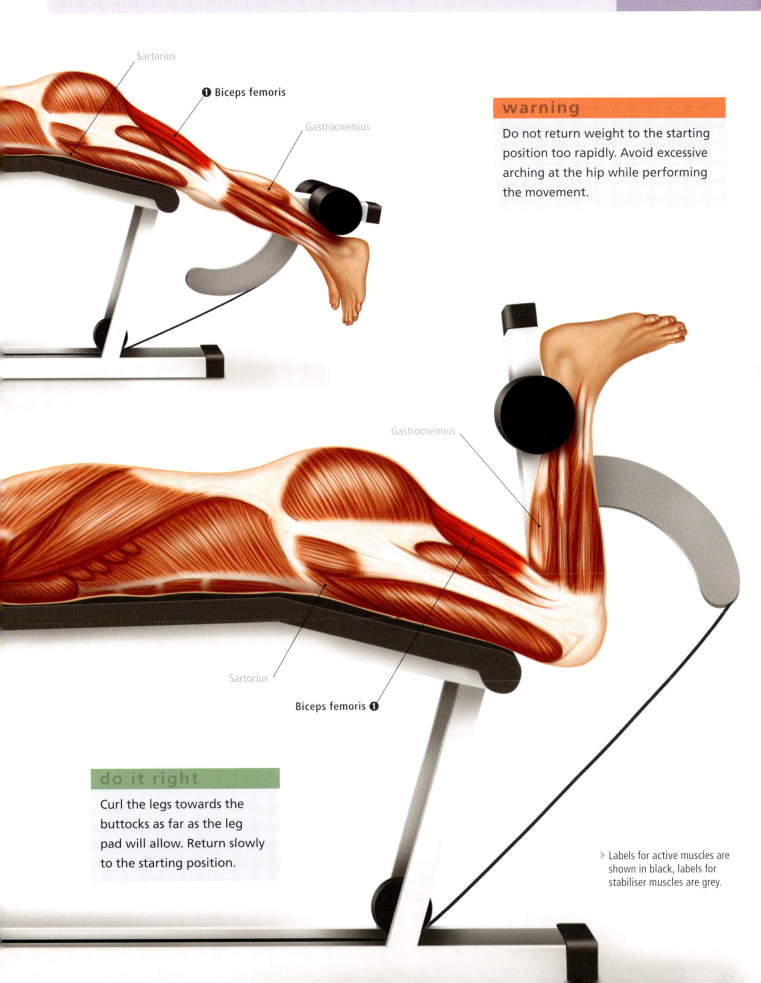

Sartorius

❶ Biceps femoris

Gastrocnemius

Gastrocnemius

Sartorius

Biceps femoris ❶

do it right

Curl the legs towards the buttocks as far as the leg pad will allow. Return slowly to the starting position.

▶ Labels for active muscles are shown in black, labels for stabiliser muscles are grey.

Leg Press

This standard exercise is a common choice for targeting the quadriceps and gluteus maximus muscles. It is a popular alternative for trainers who prefer not to use the squat exercise. The leg press is beneficial to any sports that involve running, jumping, and kicking. It has both seated and lying variations, and is suitable for both beginners and advanced trainers. Although the leg press requires a relatively expensive machine, it does permit the use of heavy weights for more advanced lifters. For beginners, the focus is on using correct technique. The following explanation centres on the 45-degree leg press, which is a standard piece of equipment in most gyms.

1

Vastus lateralis ❺
Vastus medialis ❻

Adductor longus ❸

Hamstrings

Gluteus maximus ❼

2

how to

Assume a seated, reclined position and place feet on the baseplate above, toes pointing forwards. Apply pressure to the baseplate, to 'take' the weight, then release the safety catches. Slowly lower the weight towards the body as far as is comfortable. Push legs back out until fully extended. After completing the desired number of repetitions, replace safety catches and slowly relax the pressure on the baseplate.

variations

EASY

Perform the standard exercise, but only lower the baseplate through the initial quarter of the range of motion. As strength builds, be sure to progress to the standard exercise because this variation does not train the muscles through the full range of motion.

HARD

Choose a weight that is light enough to slowly lower with one leg in a controlled movement, but heavy enough that it requires both legs to raise it again. The enhanced eccentric load strengthens muscles by increasing the stress on the muscle fibres to a greater level than is possible if both legs perform the lowering movement together. Perform sets on each leg.

active muscles

❶ Adductor brevis (under Adductor longus)
❷ Vastus intermedius (under Rectus femoris)
❸ Adductor longus
❹ Adductor magnus
❺ Vastus lateralis
❻ Vastus medialis
❼ Gluteus maximus

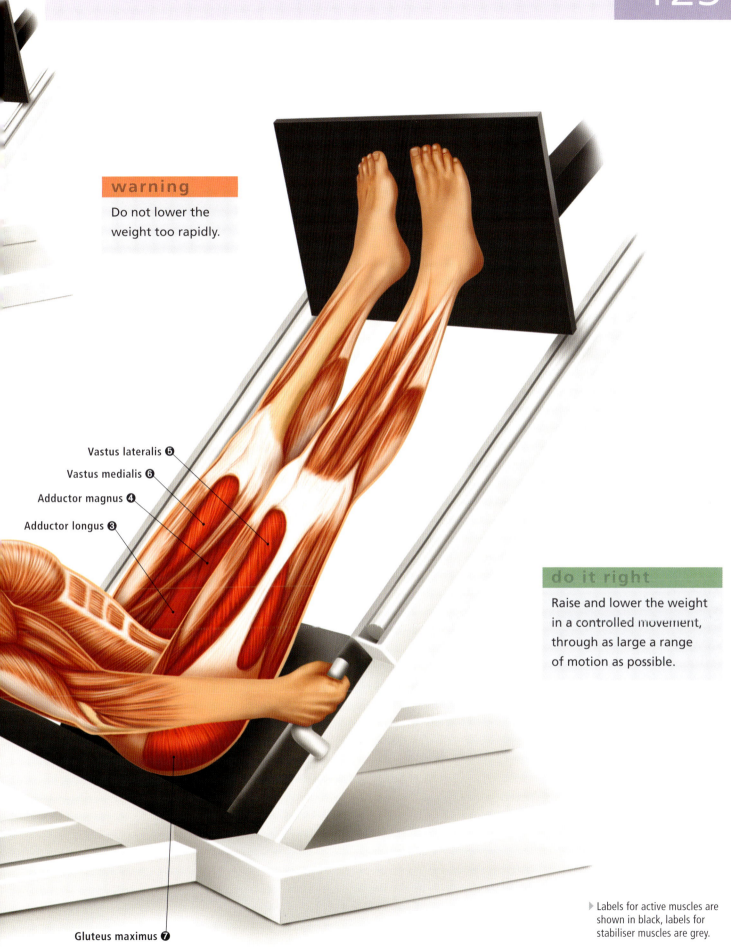

warning

Do not lower the
weight too rapidly.

Vastus lateralis ❺
Vastus medialis ❻
Adductor magnus ❹
Adductor longus ❸

do it right

Raise and lower the weight
in a controlled movement,
through as large a range
of motion as possible.

Gluteus maximus ❼

▶ Labels for active muscles are
shown in black, labels for
stabiliser muscles are grey.

Nordic Hamstrings

This hamstring exercise is specifically aimed at developing eccentric strength of the hamstring muscle group. Use it for both general strengthening programmes and in rehabilitation settings. The Nordic hamstring exercise is also a popular choice for sports that have a high incidence of hamstring injuries – the main culprits being football and track sprinting, because of the eccentric contribution just prior to and during foot strike in the running motion. It is suitable for beginners through to advanced exercisers – match the resisted range of motion and external weight to the training status of the individual.

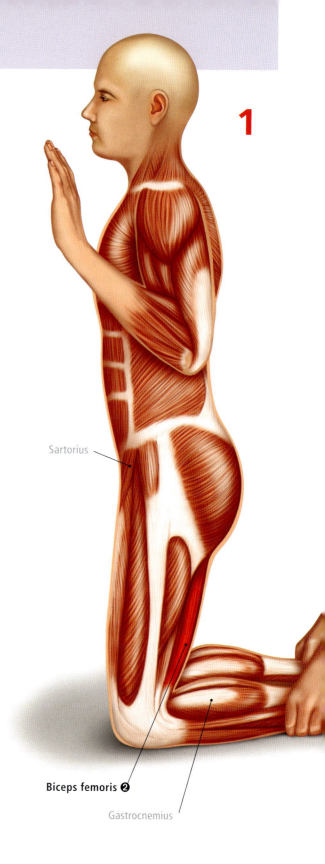

1

Sartorius

Biceps femoris **❷**

Gastrocnemius

how to

Kneel on a flat, padded surface, such as a sit-up mat, with back straight, hands at chest level, and palms facing away from the body. Have a spotter hold the lower legs down. Slowly lower body towards the ground in a controlled movement, resisting the urge to fall by contracting the hamstring muscles. Do not bend at the hips. At ground level, use a push-up motion to return to the starting position. Perform three to five repetitions per set.

variations

EASY
Resist with the hamstrings only until the controlled downwards motion cannot be contained. At that point, relax hamstrings and fall freely to the ground, slowing the descent with arms and hands. Complete the desired number of repetitions.

HARD
Add extra weight to the body by holding a weight plate or dumbbell across the chest, then perform as per the standard exercise. The extra weight increases the stress on the hamstring muscles, allowing progressive overload to be incorporated into the training programme.

active muscles

❶ Semitendinosus
❷ Biceps femoris
❸ Semimembranosus

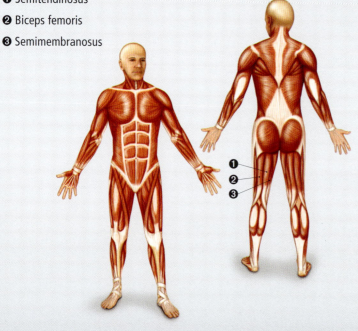

warning

If experiencing excessive stress on hamstring muscles, relax at that point and use arms and hands to break the fall.

2

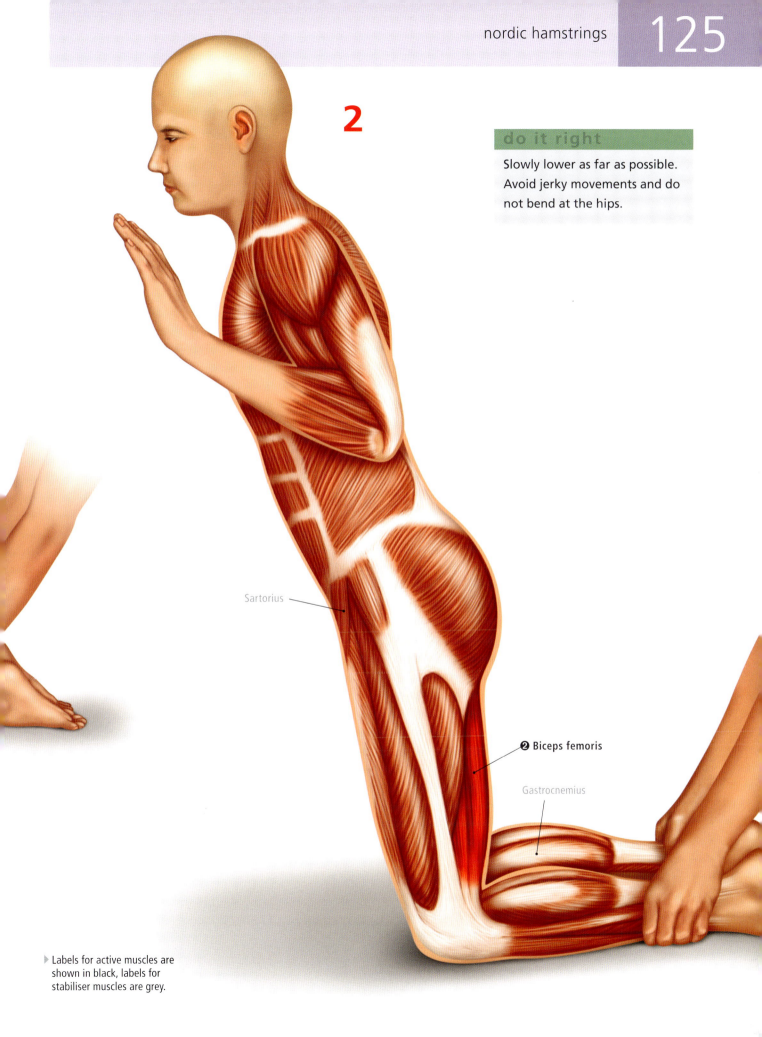

do it right

Slowly lower as far as possible. Avoid jerky movements and do not bend at the hips.

Sartorius

❷ Biceps femoris

Gastrocnemius

▷ Labels for active muscles are shown in black, labels for stabiliser muscles are grey.

Hip Adduction

This strengthening exercise is too often ignored, especially by male trainers. It provides a great workout for the adductor muscles, which pull the legs together (adduction) and allow flexion, extension, and medial or lateral rotation at the hip. This makes it a clever choice for football and many court sports, which involve cutting manoeuvres and crossover steps. Depending on the variation chosen, it can be suitable for both home and gym environments and is appropriate for beginners through to advanced trainers. The adductors are not strong muscles, so choose a light weight when first performing this exercise and progressively overload as muscles become stronger.

how to	Sit in a hip-adduction machine, with padded supports lying along the inside of the legs. Choose and set a starting position with legs open as wide as comfortably possible. Pull the padded supports together until they touch, by pushing against them with the legs. Slowly open legs, resisting against the pull of the weight return to the starting position.
EASY	Lie on the ground and push one leg at a time against a stability ball that is fixed against an immovable object, such as the other leg or a wall. Keeping the leg straight, try to depress the ball as much as possible. Slowly resist the expansion of the ball as the leg moves outwards to return to the starting position. Repeat on alternate leg.
HARD	Stand side-on to a standing pin-loaded weight stack, legs 91cm (3ft) apart. Attach a low pulley strap just above the ankle of the leg closest to the machine. The leg should be straight and raised to the side, 15cm (6 inches) off the ground, with the other leg firmly on the floor. Slowly pull the raised leg towards the ground-based leg until they meet. Return to the starting position. Repeat on alternate leg.

variations

active muscles

❶ Pectineus

❷ Adductor brevis
(under Adductor longus)

❸ Adductor longus

❹ Adductor magnus

❺ Gracilis

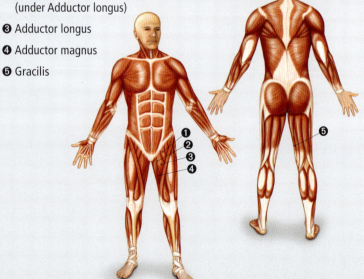

do it right

Control the speed of inwards and outwards movements. Keep legs straight while performing the exercise.

warning

Do not use excessive weight, as adductor muscles cannot tolerate the same loads as muscles such as the quadriceps.

▶ Labels for active muscles are shown in black, labels for stabiliser muscles are grey.

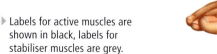

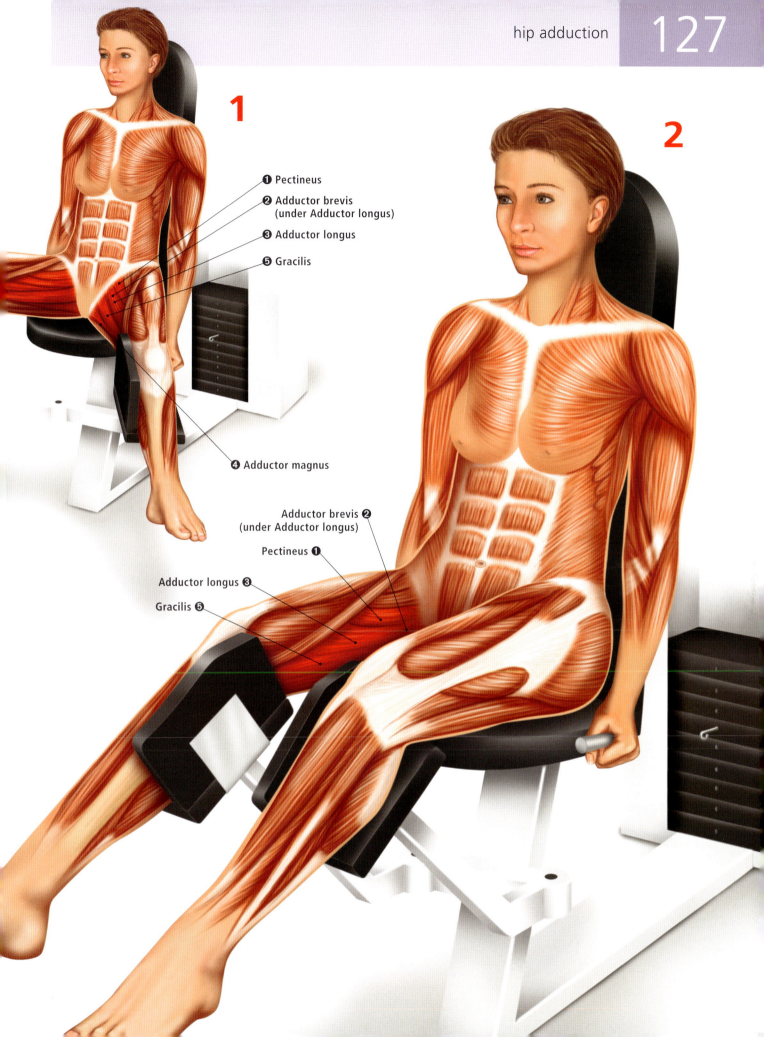

1

2

❶ Pectineus
❷ Adductor brevis
(under Adductor longus)
❸ Adductor longus
❺ Gracilis

❹ Adductor magnus

Adductor brevis ❷
(under Adductor longus)

Pectineus ❶

Adductor longus ❸

Gracilis ❺

Hip Abduction

This often-overlooked exercise, especially by male trainers, is effective in working the hip abductor muscles that move the legs apart. It is a good choice to balance exercise programmes, as it improves weakness of the gluteus medius muscles. This also makes it a popular choice for sports-specific training and occupational rehabilitation programmes. The abductors are not a strong muscle group, so take care choosing weight when starting to train the area. This exercise is suitable for home and gym environments, depending on the variation, and is appropriate for all levels of trainers.

how to

Sit in the hip-abduction machine, with the padded supports lying along the outside of the legs. Choose and set a starting position with legs closed. Push the padded supports apart as far as possible, then resist against the weight as legs slowly close to return to the starting position. Complete the desired number of repetitions.

variations

EASY

Place a stability ball against an immovable surface, such as a wall. Stand or lie with one leg against the ball and feet close together. Hold the supporting leg still, and apply pressure to the ball with the other leg. Squash the ball against the wall as legs open, keeping both legs straight. During the return phase of the motion, slowly resist the expansion of the ball as legs are pushed back together. Repeat on alternate leg.

HARD

Stand side-on to a standing pin-loaded weight stack. Attach a low pulley strap just above the ankle of the leg furthest from the machine. Both legs should be straight, feet on the ground. Slowly raise the leg with the attached strap outwards, away from the body, until the foot is about 61 cm (2 ft) off the ground. Slowly return foot to the ground, resisting the pull of the weight. Repeat on alternate leg.

active muscles

❶ Tensor fasciae latae
❷ Piriformis
❸ Gluteus medius
❹ Gluteus minimus

(2, 3, and 4 all under Gluteus maximus)

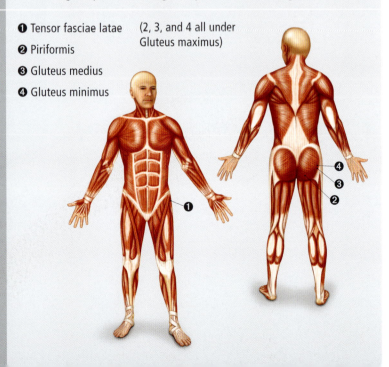

do it right

Keep legs straight throughout the exercise. Resist against the pull of the weight stack and return legs slowly to the closed position.

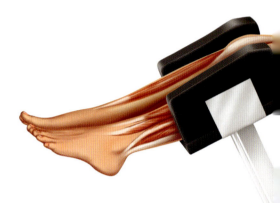

warning

Use a weight that allows a smooth, controlled motion. Maintain correct technique by avoiding any twisting of the hip.

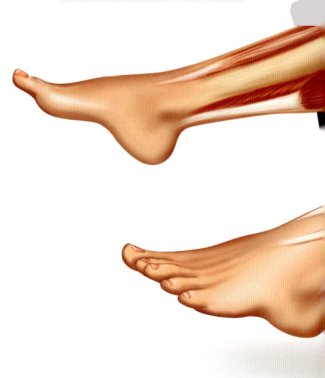

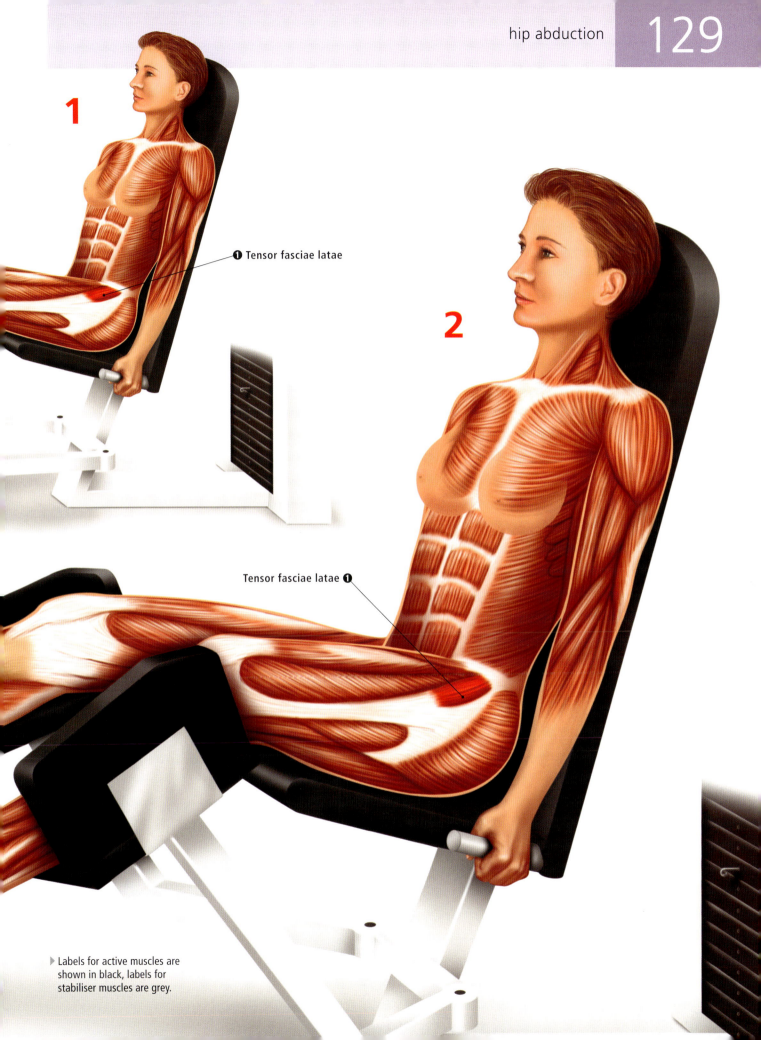

1

❶ Tensor fasciae latae

2

Tensor fasciae latae ❶

▶ Labels for active muscles are shown in black, labels for stabiliser muscles are grey.

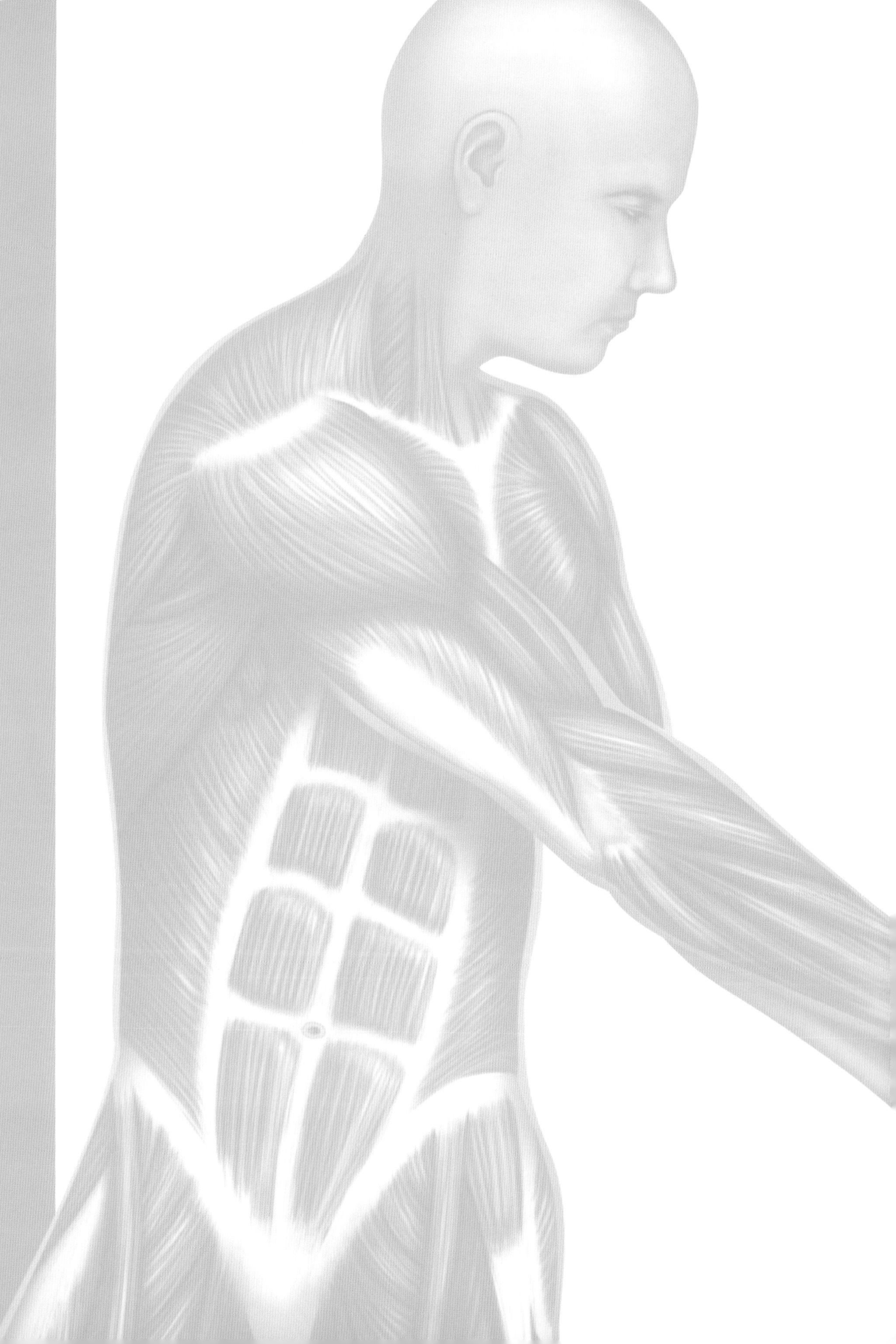

Trunk Exercises

The core, a vitally important area for strength, fitness, and stability, is the focus of exercises in this section. Do not mistake the term 'core stability' for core strength. Few people truly require strong core muscles – it is endurance, tone, and the ability to support a stable posture while completing other activities that is most important.

All exercisers will benefit from core stability and lower back endurance. Together, they protect against lower back pain – one of the most common complaints in Western society. For athletes, core stability provides a stable foundation from which to achieve longer and stronger throws, better absorb body contact, and maintain balance.

Correct technique is essential to ensure that trunk exercises are safe and effective. Maintain a neutral spine position and minimise the use of accessory muscles.

Plank

This core-stabilising exercise is also used for scapular-stabilisation training. Rather than building strength, the plank develops endurance and stability. In particular, it improves endurance in the lower back, abdominal, shoulder, and hip muscles. The aim is to hold the position for as long as possible, depending on fitness levels. The current world-record holder maintained the plank position for more than one hour! This exercise is ideal for sports where athletes need to stay stable or rigid while performing a task, such as gymnastics and diving, or sports where body contact needs to be absorbed or countered. Use the plank in early to moderate shoulder-rehabilitation programmes, and as a precursor to performing a push-up.

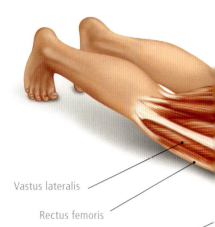

Vastus lateralis

Rectus femoris

External oblique ❸

Internal oblique ❷
(under External oblique)

Rectus abdominis ❶

how to	Start by lying face down on the ground with elbows bent, hands curled into fists just in front of the shoulders, and feet on their toes. Push the whole body up so that it rests only on the forearms and toes. Keep eyes down, shoulders back, and ensure that the back is straight. The body should be in a flat plane from the head to the heels. Hold the position for a minimum of 15 seconds. Work to increase the amount of time the position is held.
variations EASY	Keep knees on the ground rather than resting on the toes if lacking the upper body strength or endurance to hold the plank. Ensure that the body is held in a flat plane from the head to the knees.
HARD	Lift one foot off the ground for five seconds at a time. This works the oblique and gluteal muscles, as they engage to resist rotation and maintain the leg in the air. For an advanced plank workout, try the 'side plank'. Move from the left side, to centre, to the right side, holding each position for 30–60 seconds.
active muscles	❶ Rectus abdominis ❷ Internal oblique (under External oblique) ❸ External oblique

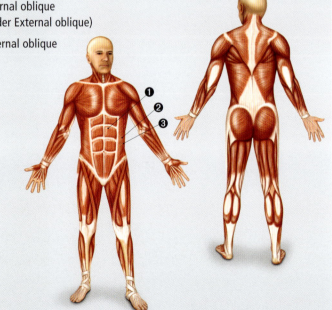

2

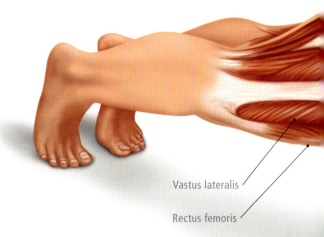

Vastus lateralis

Rectus femoris

do it right

Use a mirror when first performing this exercise to ensure correct positioning.

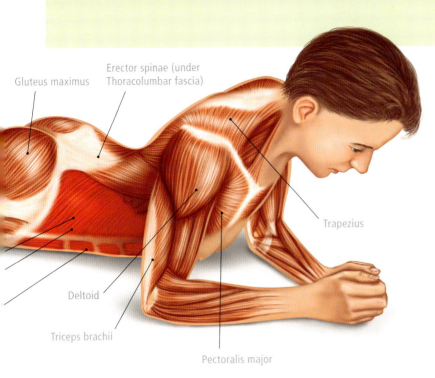

Gluteus maximus

Erector spinae (under Thoracolumbar fascia)

Trapezius

Deltoid

Triceps brachii

Pectoralis major

1

warning

Do not arch the lower back. If abdominal muscles are weak, start with the easy variation and build up endurance.

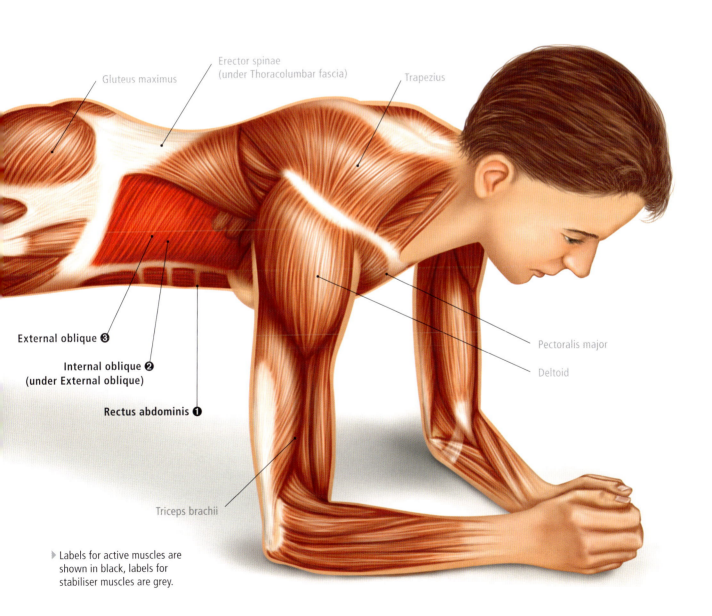

Gluteus maximus

Erector spinae (under Thoracolumbar fascia)

Trapezius

Pectoralis major

Deltoid

External oblique ❸

Internal oblique ❷ (under External oblique)

Rectus abdominis ❶

Triceps brachii

▶ Labels for active muscles are shown in black, labels for stabiliser muscles are grey.

Crunch

This classic abdominal exercise has been performed thousands of times. However, a common misconception is that it assists in losing excess fat around the girth. This is not true, as the crunch is not a fat-burning exercise; rather, it is a strength and endurance exercise for the rectus abdominis, internal and external obliques, psoas, and iliacus. In the curled position, the lower back muscles are stretched and the gluteals, quadriceps, adductors, and hamstrings help stabilise the lower half of the body. The crunch does not require any special equipment, so it can be performed just about anywhere and is great for home exercise routines or trainers on the go.

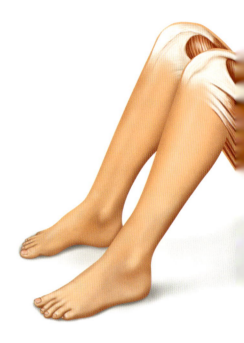

how to

Lie back, with knees and hips flexed and feet flat on the floor. Cross arms over the chest with hands flat on opposite shoulders, or for a slightly harder option, place hands lightly beside the head. Squeeze and contract abdominal muscles to lift the head, shoulders, and scapula off the floor. Slowly lower back to the starting position.

variations

EASY

Use a spotter to hold the ankles or hook the feet under a solid stationary object, such as a bench or cupboard, if abdominals do not have the strength and power needed to perform this exercise without assistance. Contract through the abdominals and do not pull up with the feet or the legs, as this reduces the effectiveness of the exercise.

HARD

Reach arms back overhead or hold a weight overhead to really engage the abdominal muscles. These variations also make the lower limb stabilisers work harder to keep the feet down on the floor. However, ensure that the correct technique is not compromised.

active muscles

❶ Rectus abdominis
❷ Internal oblique
   (under External oblique)
❸ External oblique

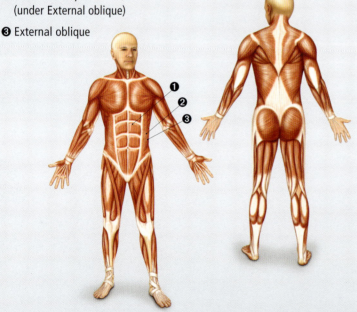

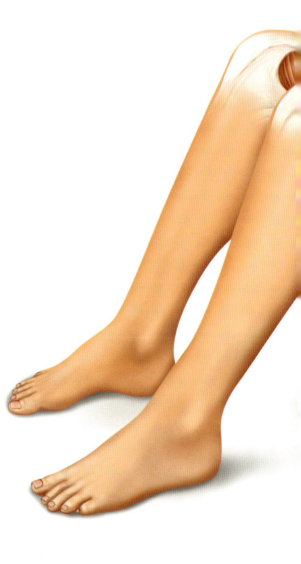

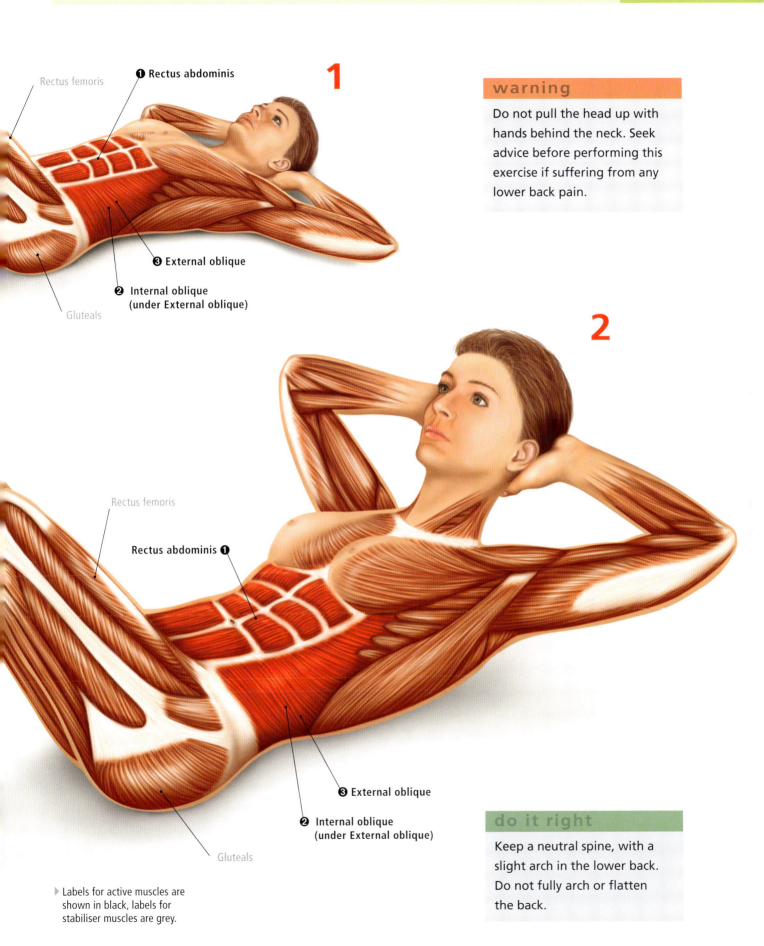

Rectus femoris

❶ Rectus abdominis

1

Rectus femoris

❸ External oblique

❷ Internal oblique
(under External oblique)

Gluteals

2

Rectus femoris

Rectus abdominis ❶

❸ External oblique

❷ Internal oblique
(under External oblique)

Gluteals

▶ Labels for active muscles are shown in black, labels for stabiliser muscles are grey.

Cross-body Crunch

This version of the exercise increases the challenge by requiring greater movement amplitude and adding a twisting component to the standard crunch. In turn, it targets a greater portion of internal and external oblique muscles and calls more on the secondary stabilisers. The cross-body crunch can be used as a progression towards the bicycle exercise, because the movement is similar but demands less from leg muscles and spinal stabilisers. Much like the standard crunch, this exercise is versatile and can be performed almost anywhere, as no special equipment is needed.

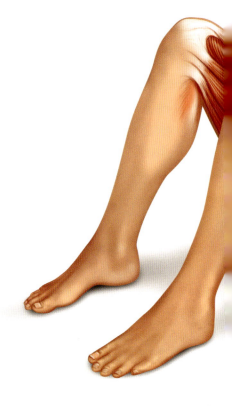

how to

Lie back with knees and hips flexed, feet flat on the floor, and hands placed lightly beside the head. Squeeze and contract the abdominals to lift the head, shoulders, and scapula off the floor while twisting towards the right side. At the same time, lift the left leg so that the right elbow and left knee touch over the chest. Slowly return to the starting position and repeat with alternate arm and leg.

variations

EASY

Drop the leg movement and just concentrate on lifting and twisting the upper half of the body if lacking the strength or endurance to execute the standard exercise in a smooth, controlled manner. This variation minimises the abdominal work and decreases the importance of secondary stabilisers.

HARD

Perform the exercise on an incline bench, with head lower than the feet. This position increases the resistance by simply adding an extra gravitational pull. To increase the resistance even further, hold a weight behind the head.

do it right

Be sure to feel the 'squeeze' when contracting the abdominals. Do not pull up with the arms or legs.

active muscles

❶ Rectus abdominis

❷ Internal oblique
   (under External oblique)

❸ External oblique

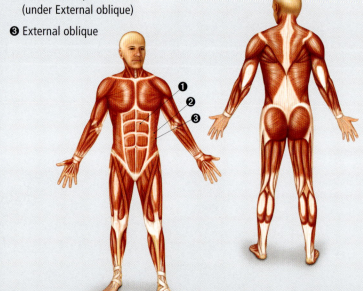

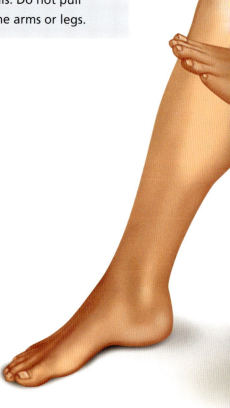

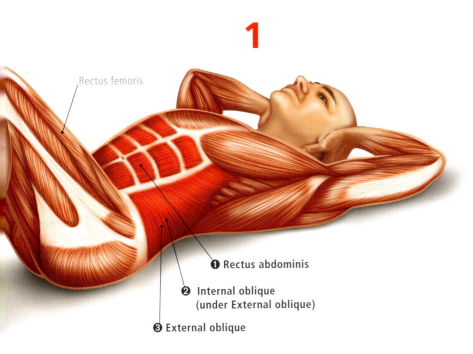

1

Rectus femoris

❶ Rectus abdominis

❷ Internal oblique
(under External oblique)

❸ External oblique

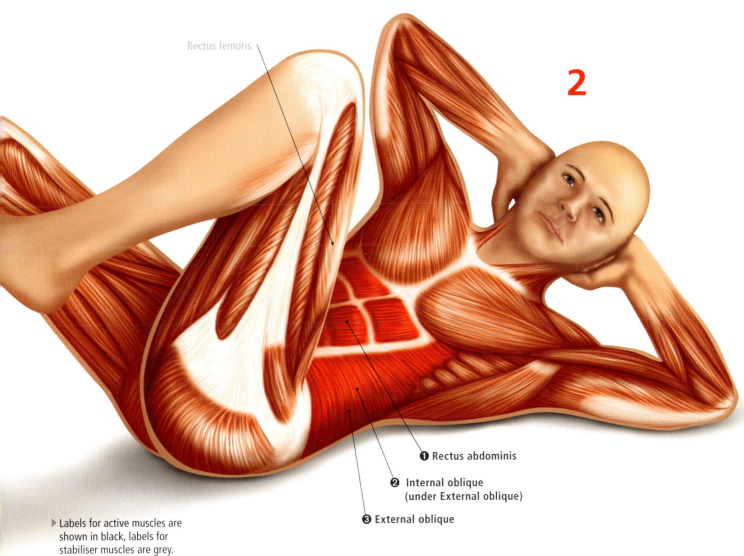

Rectus femoris

2

❶ Rectus abdominis

❷ Internal oblique
(under External oblique)

❸ External oblique

▶ Labels for active muscles are shown in black, labels for stabiliser muscles are grey.

Bicycle

This advanced exercise works the rectus abdominis hard. In fact, in a San Diego State University study that measured muscle activation, it ranked top of all common abdominal exercises for rectus activation. The bicycle also ranked highly for activation of the obliques, making it one of the most effective ways to target the superficial abdominal muscles. Definition in these muscles results in the 'six-pack' look that so many trainers desire. Use the bicycle as an advanced core-stability exercise, but only once deep core muscles have been effectively trained. Maintaining a neutral spine while lifting and moving the legs places high demands on the muscles.

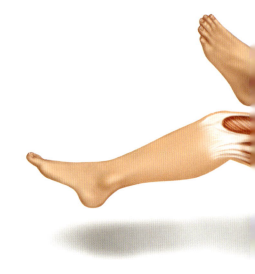

how to	Lie flat on the floor, with the lower back pressed to the ground, then contract the core muscles. Holding the head gently with the hands, lift the knees to a 45-degree angle. Move the legs through a bicycle pedal motion, and alternately touch the elbow to the opposite knee while twisting back and forth. Breathe evenly throughout the exercise.
variations **EASY**	The most challenging part of the bicycle is maintaining stability when the leg is fully extended. Keep knees slightly flexed throughout the exercise if finding it hard to maintain correct form or a neutral spine position.
HARD	Try a 'seated bicycle'. Sit down and lean back slightly to make the core stabilisers start working even before any movement begins. Extend legs and lift both heels off the floor. Perform the bicycle pedal action and move opposite elbows to knees. To make this exercise even more challenging, touch the outside of the elbow to the outside of the knee.
active muscles	❶ Rectus abdominis ❷ Internal oblique (under External oblique) ❸ External oblique

do it right

Throughout the exercise, pull the deep abdominal muscles towards the spine and maintain a neutral spine position.

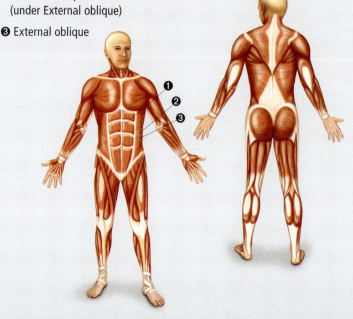

▶ Labels for active muscles are shown in black, labels for stabiliser muscles are grey.

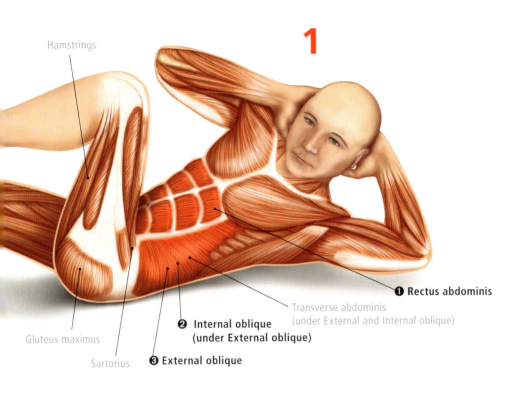

1

Hamstrings

Gluteus maximus

Sartorius

❶ **Rectus abdominis**

Transverse abdominis
(under External and Internal oblique)

❷ **Internal oblique**
(under External oblique)

❸ **External oblique**

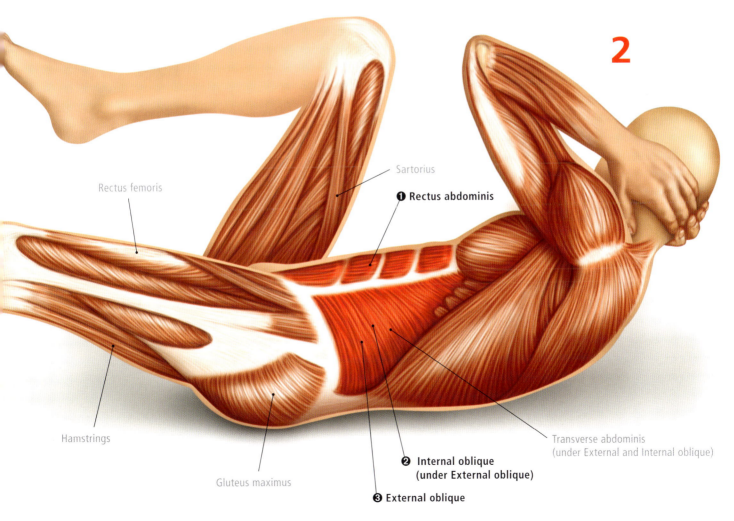

2

Rectus femoris

Sartorius

❶ **Rectus abdominis**

Hamstrings

Gluteus maximus

❷ **Internal oblique**
(under External oblique)

❸ **External oblique**

Transverse abdominis
(under External and Internal oblique)

Superman

This core-stability exercise is also known as the 'bird dog', as the position moves between resting on all fours and extending arms and legs. It works predominantly on the posterior core, which is made up of the obliques, erector spinae, multifidus, and gluteals. The superman develops strength and stability around the neck and shoulder girdle by improving extensor endurance and encouraging rotational stability. Good lumbar extensor endurance is essential to protecting against lower back pain, and rotational stability is important in many sports, especially bat and racket sports. The superman does not require any special equipment – it is a floor or mat exercise and, as such, can easily be part of a home or gym routine.

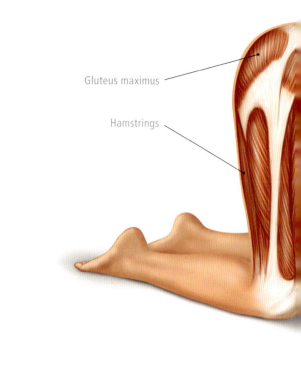

Gluteus maximus

Hamstrings

how to	Kneel on the ground with hands placed in front of the body, about shoulder-width apart. Tilt the pelvis back and forth to find the neutral spine position. In this position, brace the abdominals and lift one hand and the opposite knee just off the ground, balancing on the alternate hand and knee. Once comfortably stable, fully extend the arm and leg. Try to keep a flat plane all the way from the hand to the foot. Hold this position for about ten seconds. Return to the starting position and extend on the opposite side.

| variations | EASY | Keep both hands on the ground and extend only one leg at a time if finding it hard to stabilise with both leg and arm extended. |
| | HARD | Start on hands and toes, rather than on hands and knees, for an advanced workout. Shoulder stabilisers, core stabilisers, and hamstrings will all have to work harder because of the longer segment being stabilised and the decreased surface contact area in this variation. |

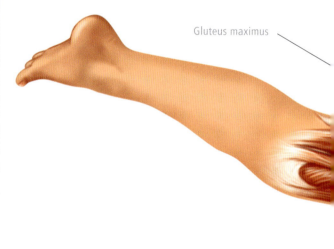

Gluteus maximus

Hamstrings

active muscles

❶ Rectus abdominis

❷ Internal oblique (under External oblique)

❸ External oblique

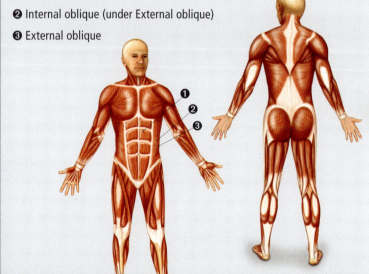

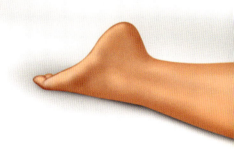

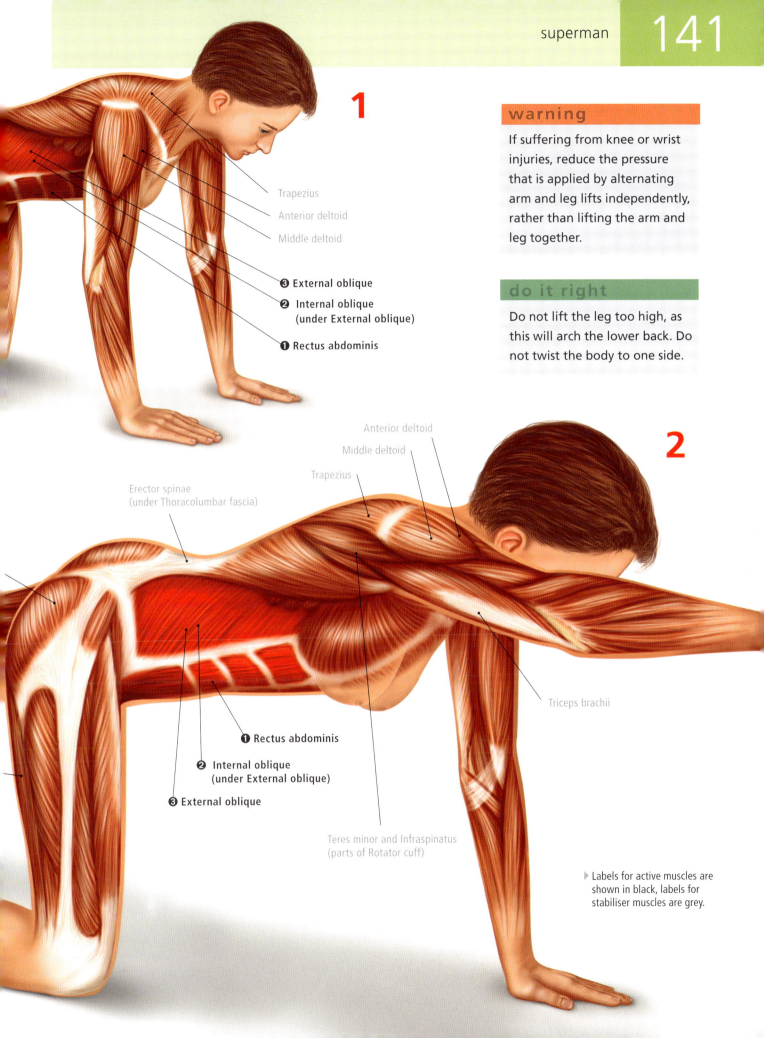

1

Trapezius

Anterior deltoid

Middle deltoid

❸ External oblique

❷ Internal oblique
(under External oblique)

❶ Rectus abdominis

2

Anterior deltoid

Middle deltoid

Trapezius

Erector spinae
(under Thoracolumbar fascia)

Triceps brachii

❶ Rectus abdominis

❷ Internal oblique
(under External oblique)

❸ External oblique

Teres minor and Infraspinatus
(parts of Rotator cuff)

▶ Labels for active muscles are shown in black, labels for stabiliser muscles are grey.

Bridge

This strengthening exercise targets the gluteal, hamstring, abdominal, and lower back muscles. It is an excellent rehabilitation exercise for lower back pain and core stability. Poor endurance of the lower back muscles is one of the most dangerous risk factors in developing lower back pain. In response, the bridge has proven to be one of the most effective ways to activate and strengthen lower back muscles, such as the multifidus and the erector spinae. Use this exercise as a preventative measure, or as part of a training programmeme when recovering from injury. There are many variations of the bridge, so exercisers can progress from easy to advanced forms. It requires little equipment and is therefore easy to perform in most settings.

1

how to	Lie back with hands by sides, knees bent, and feet flat on the floor. Tighten the abdominal and gluteal muscles, push down through the heels, and raise the hips to create a straight plane from the knees to the shoulders. Keep the core muscles tight and do not let hips drop or the back arch. Hold this position for 20–30 seconds. Once the muscles begin to fatigue, slowly lower back to the starting position.

warning

Dropping suddenly from the raised position can jar the lower back and cause pain or injury.

variations	**EASY**	Depending on individual fitness levels, lift the hips only a small way off the ground until body strength develops, or hold the standard position for a reduced amount of time.
	HARD	Try a 'one-legged bridge'. Perform the exercise as outlined in the standard version but start with only one knee bent, holding the other leg fully extended. Push up with the bent leg and aim for a flat plane from the shoulders to the foot of the straight leg. Resist the urge to rotate the body; keep hips flat and do not twist to one side.

active muscles

❶ Erector spinae

❷ Multifidus (under Erector spinae)

❸ Rectus abdominis

❹ Internal oblique (under External oblique)

❺ External oblique

❻ Gluteus maximus

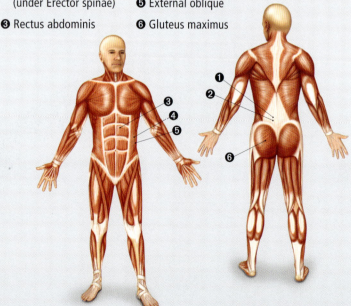

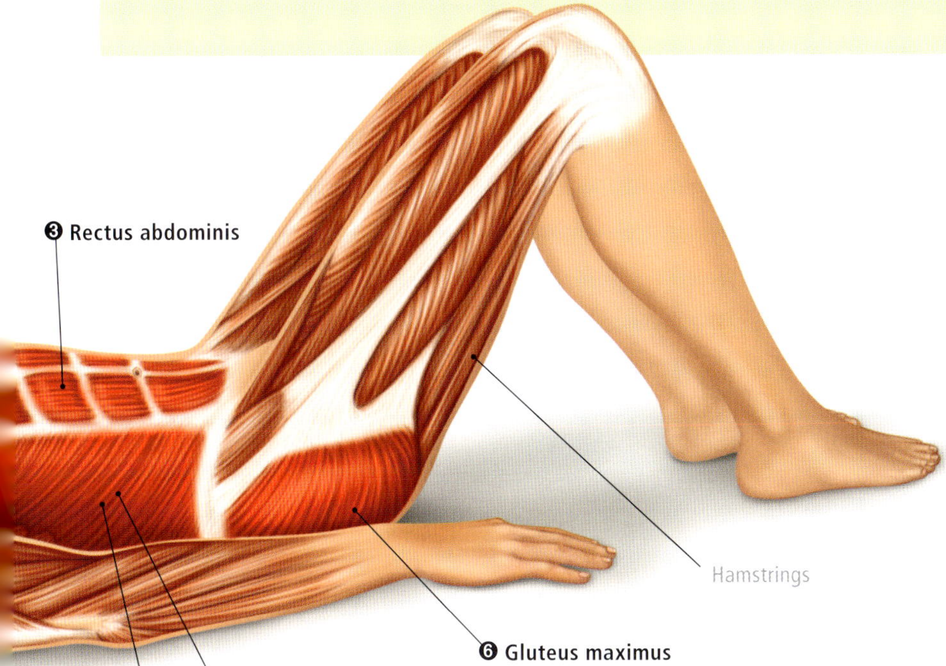

❸ Rectus abdominis

Hamstrings

❻ Gluteus maximus

❹ Internal oblique
(under External oblique)

❺ External oblique

do it right

Keep the core tight by pulling the belly button in towards the spine.

2

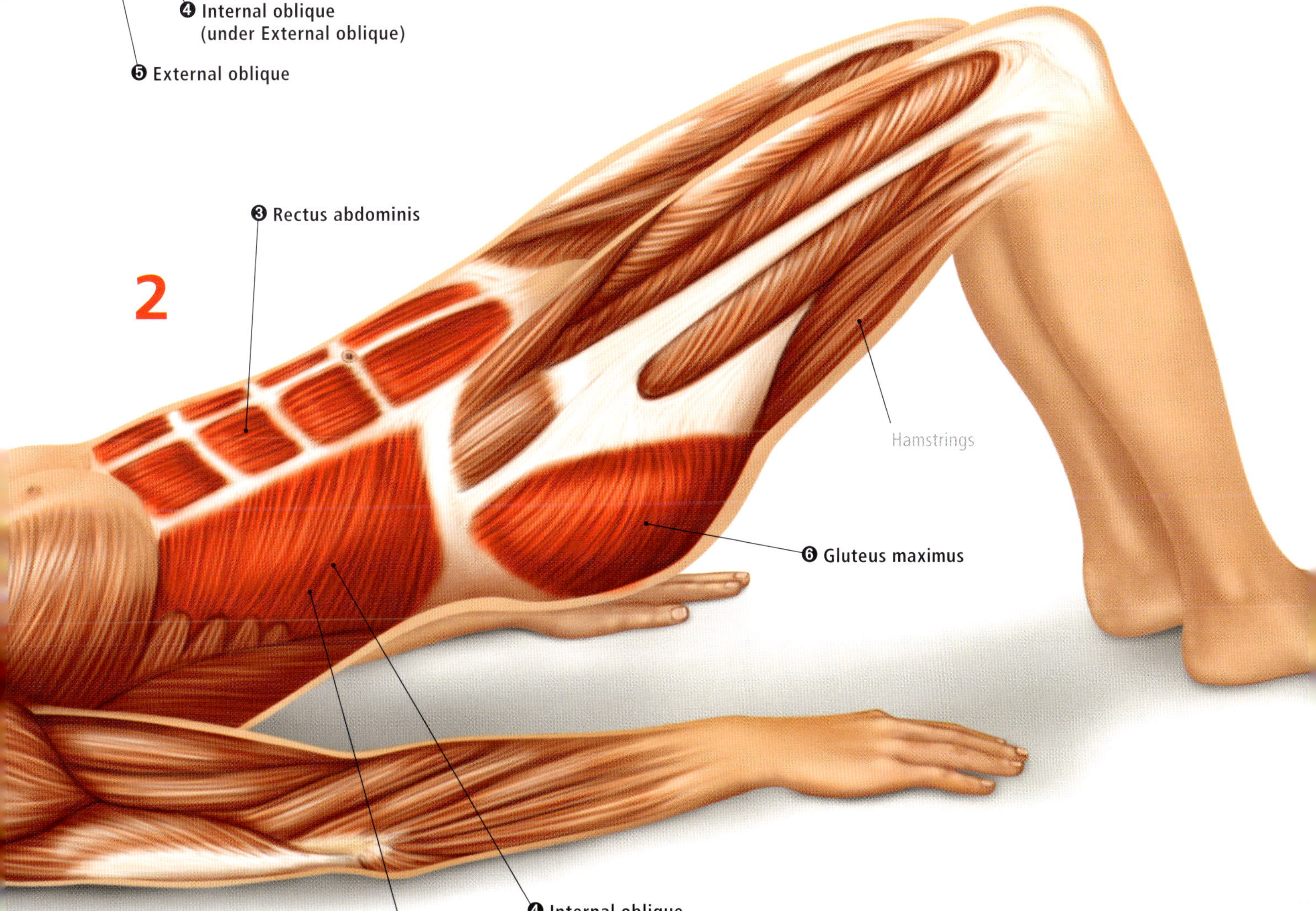

❸ Rectus abdominis

Hamstrings

❻ Gluteus maximus

❹ Internal oblique
(under External oblique)

❺ External oblique

▶ Labels for active muscles are shown in black, labels for stabiliser muscles are grey.

Back Extension

This bodyweight exercise both stretches and strengthens the muscles of the lower back. It is usually performed in the gym on a back-extension machine. In the lowered position, erector spinae, quadratus lumborum, and the thoracolumbar fascia are stretched. Then, as the body is raised, the back muscles and hamstrings must contract. Use the back extension to complement crunches or sit-ups and balance the mid-section. Many people perform abdominal exercises and upper back exercises but neglect the muscles in the lower back. Remember, strength and endurance in these muscles are essential to protect against developing lower back pain.

warning

Even the easy variation of this exercise puts strain on the lower back. Discontinue if experiencing any pain while performing the movement.

how to

Use a back-extension machine and position the pads to sit underneath the thighs and on top of the calves. Adjust the length to support the hips and pelvis, but ensure that none of the upper body is supported. Start in a neutral or slightly hyperextended position, then lower the body, bending at the waist. Lower as far down as back and hamstring flexibility will allow. Lift up and return to the starting position.

variations

EASY

Lie face down on the floor instead of using a back-extension machine. Place hands behind the back and arch up to lift the head, neck, shoulders, and chest off the ground. This variation still works through the same range of muscles as the standard form, without relying on a back-extension machine.

HARD

Once comfortably able to complete three sets of 10–12 repetitions lifting your own bodyweight, try a weighted back extension. Take a plate from the barbell racks and hold it across your chest as you perform the movement. For safety, place the plate down on the ground at the end of the set, before dismounting the machine.

active muscles

❶ Erector spinae
❷ Multifidus
❸ Quadratus lumborum

(1, 2, and 3 all under Thoracolumbar fascia)

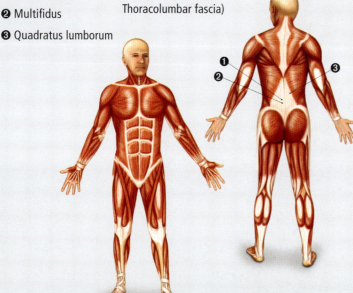

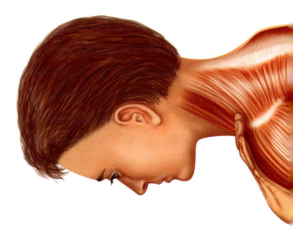

2

do it right

Do not swing or bounce at end positions; use a continuous, smooth motion.

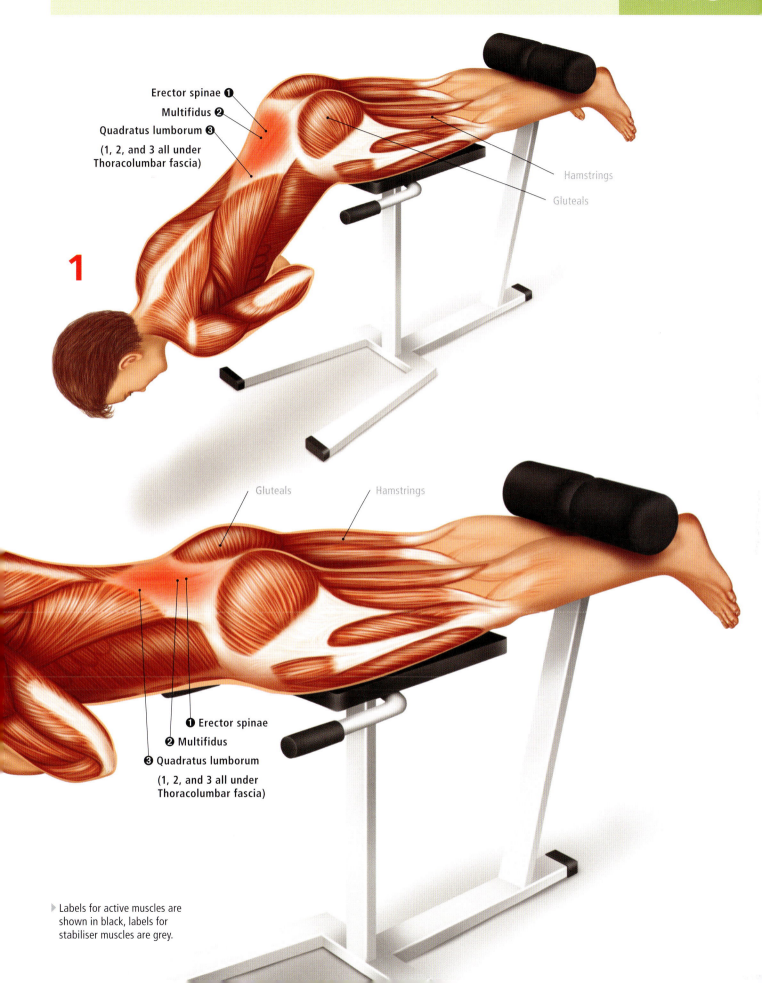

Erector spinae ❶
Multifidus ❷
Quadratus lumborum ❸

(1, 2, and 3 all under
Thoracolumbar fascia)

1

Hamstrings

Gluteals

Gluteals

Hamstrings

❶ Erector spinae
❷ Multifidus
❸ Quadratus lumborum

(1, 2, and 3 all under
Thoracolumbar fascia)

▷ Labels for active muscles are
shown in black, labels for
stabiliser muscles are grey.

Lift

This proprioceptive neuromuscular facilitation (PNF) exercise incorporates complex multidirectional movements that involve both the upper and lower limbs. It is often used in conjunction with the chop. The PNF lift comprises extension and rotation movements that strengthen torso rotation, the upper back, chest, shoulders, and arms. It can be performed with cable-weight machines, elastic-resistance bands, or a medicine ball. As with all core stability exercises, the aim is not to lift the heaviest weight, but to maintain perfect form throughout the set. Add the lift to later stages of injury rehabilitation programmes and use it for sports-specific training. This exercise is perfect for racket and bat sports, especially golf.

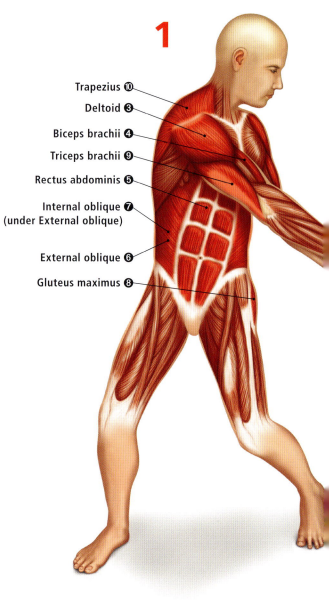

1

Trapezius ❿
Deltoid ❸
Biceps brachii ❹
Triceps brachii ❾
Rectus abdominis ❺
Internal oblique ❼
(under External oblique)
External oblique ❻
Gluteus maximus ❽

how to

The starting position can vary from kneeling, half kneeling, sitting on a stability ball, or standing. However, the movement pattern is always the same: from low to high; lifting with both hands; beginning with a pulling motion that moves across the midline of the body and ending with a pushing motion. For a standing lift, stand side-on or slightly forwards and grasp the cable handle with both hands on one side of the body, below knee height. Lift with the upper or crossed-over arm dominating, twist and extend with the trunk, then push through and up with the other arm so that the movement ends on the other side of the body, hands above shoulder height. Slowly lower to the starting position. Perform sets from right to left, then left to right.

variations

EASY
Stand with a wide or split stance to provide a greater base of support for the body and achieve more stability from the legs.

HARD
Stand on one leg to decrease the stability supplied by the legs and put the entire focus of the exercise on the core muscles.

active muscles

❶ Erector spinae
❷ Multifidus (under Erector spinae)
❸ Deltoid
❹ Biceps brachii
❺ Rectus abdominis
❻ External oblique
❼ Internal oblique (under External oblique)

❽ Gluteus maximus
❾ Triceps brachii
❿ Trapezius

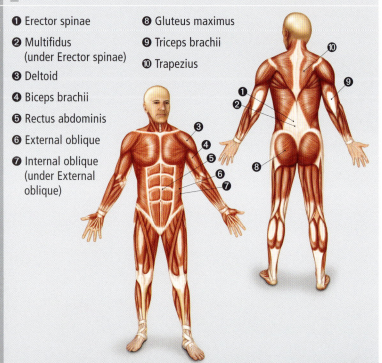

warning

Do not try to progress this exercise too quickly. It is all about form and technique. Too much weight too quickly can lead to injury.

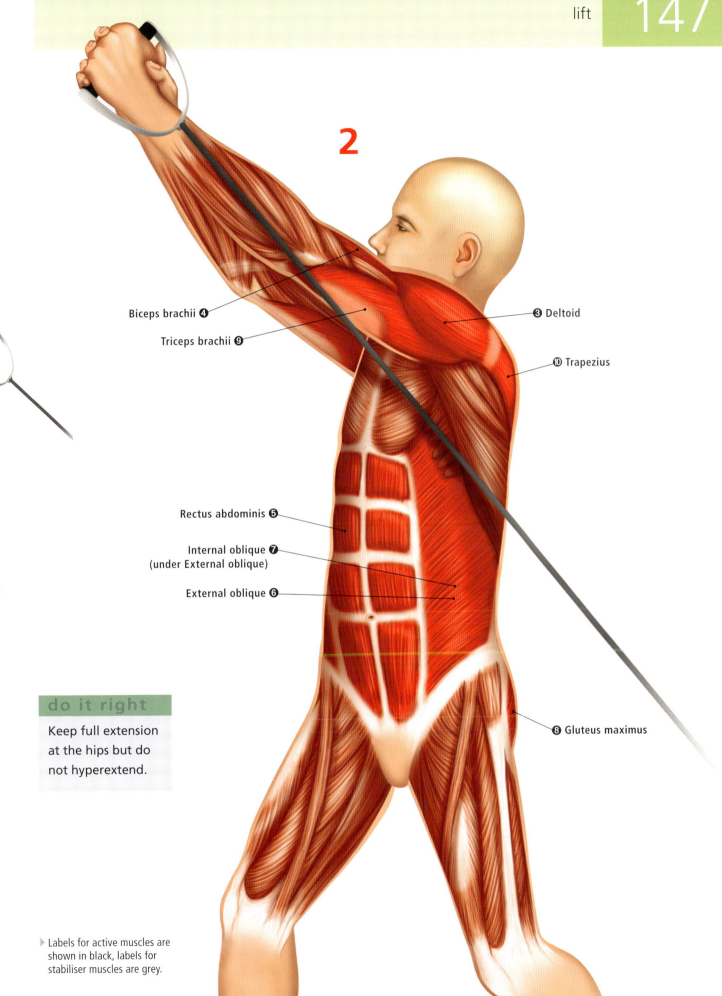

2

Biceps brachii ❹

Triceps brachii ❾

❸ Deltoid

❿ Trapezius

Rectus abdominis ❺

Internal oblique ❼
(under External oblique)

External oblique ❻

❽ Gluteus maximus

do it right
Keep full extension
at the hips but do
not hyperextend.

▶ Labels for active muscles are
shown in black, labels for
stabiliser muscles are grey.

Chop

This companion exercise to the lift is essentially a mirror image of that exercise, this time moving from a high to a low position. Because of the mechanical advantage that the chop has, most exercisers will be able to lift about one-third more weight in this exercise compared to the lift. When combined with left and right, the chop-lift combination targets four core quadrants. A skilled physical therapist can specifically test for strength in each quadrant, to determine whether a client should target one area first or train all four immediately.

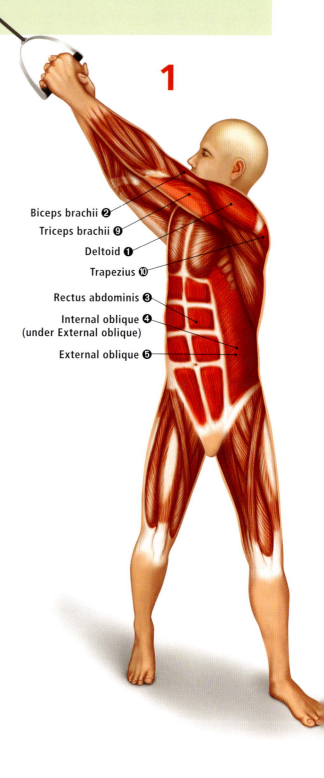

1

Biceps brachii ❷
Triceps brachii ❾
Deltoid ❶
Trapezius ❿
Rectus abdominis ❸
Internal oblique ❹
(under External oblique)
External oblique ❺

how to

The starting position can vary between kneeling, sitting, and standing. Grasp the handle with one palm facing down and one palm facing up. The chop movement is always from high to low, beginning with a pulling motion, where the crossed-over arm moves across the mid-line of the body, and ending with a pushing motion from the following arm. Choose a starting weight that will cause fatigue or loss of the stable position after 6–12 repetitions. Try to correct the position and to keep going when starting to lose the stable base. If unable to maintain form and posture, then finish the set. Repeat from the opposite side.

variations

EASY

Stand with feet wide apart and knees slightly bent. The greater the stability supplied by the legs, the less that is required from the core. Use elastic tubing to perform this exercise at home or on the road, to continue resistance training when unable to get to the gym.

HARD

Kneel on a stability ball for a real challenge! Using the ball incorporates an unstable surface, and by adding rotational movements, only those with extremely high levels of core stability will be able to keep this variation under control.

active muscles

❶ Deltoid
❷ Biceps brachii
❸ Rectus abdominis
❹ Internal oblique (under External oblique)
❺ External oblique
❻ Gluteus maximus
❼ Multifidus (under Erector spinae)
❽ Erector spinae
❾ Triceps brachii
❿ Trapezius

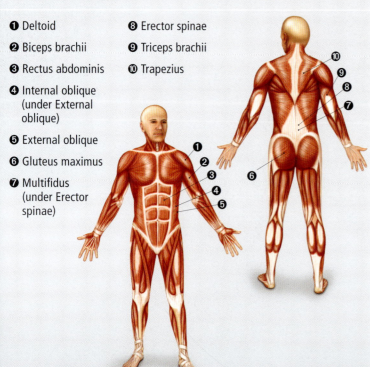

warning

Do not kneel or stand on a stability ball unless well practised and perfectly comfortable with this position.

2

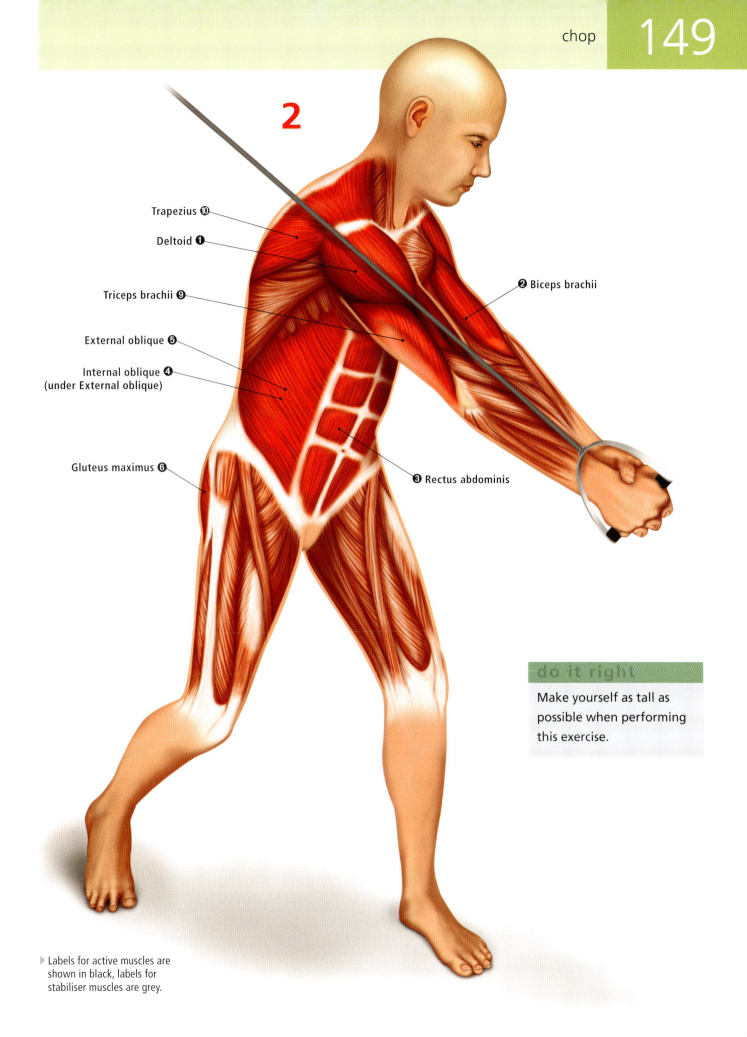

Trapezius ⑩

Deltoid ❶

Triceps brachii ❾

External oblique ❺

Internal oblique ❹
(under External oblique)

Gluteus maximus ❻

❷ Biceps brachii

❸ Rectus abdominis

Make yourself as tall as possible when performing this exercise.

▶ Labels for active muscles are shown in black, labels for stabiliser muscles are grey.

Walkout

This stability-ball exercise works both the abdominal and back muscles. Using a stability ball forces multiple muscle groups to activate as the body struggles to remain balanced on the ball, and is a simple way to increase the challenge of many exercises. During the walkout, the abdominal and lower back muscles are active isometrically, while the shoulder and scapular stabilisers must maintain stability at the glenohumeral joint. Include this exercise to train core stability or rigidity, as well as to strengthen and stabilise the shoulder capsule. It is suitable as a general toning exercise, part of sport-specific training, or as an element of a rehabilitation programme for back, neck, or shoulder injuries.

warning

Move to more challenging forms only when able to keep the ball completely stable in the full walkout position.

do it right

Do not let the lower back or hips sag when walking out. Keep the body rigid like a plank.

Hamstrings

Quadriceps femoris

2

how to

Lie on the stomach over a stability ball so that hands and toes can reach the floor. Place hands on the ball underneath the shoulders and contract the abdominal muscles. Lift feet off the floor and extend the legs so that the torso and legs are in a flat, horizontal plane. Keep the legs rigid and slowly walk hands forwards. The ball will roll slightly, but keep walking out until only the feet rest on top of the ball. Slowly walk back and return to the starting position, with the stomach resting on top of the ball.

variations

EASY

The challenge of this exercise increases the further an exerciser walks out over the ball. Stop when thighs or knees are on top the ball if unable to keep stable walking all the way out to the feet.

HARD

For an extra stability challenge, lift one foot off the ball at the end of the walkout. Keep legs fully extended and do not let the ball roll from side-to-side underneath the foot. For an extra strength challenge, add a push-up when in the full walkout position. To increase the challenge even further, combine both variations.

active muscles

❶ Erector spinae
   (under Thoracolumbar fascia)

❷ Rectus abdominis

❸ Internal oblique
   (under External oblique)

❹ External oblique

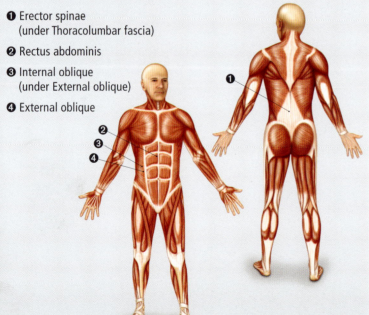

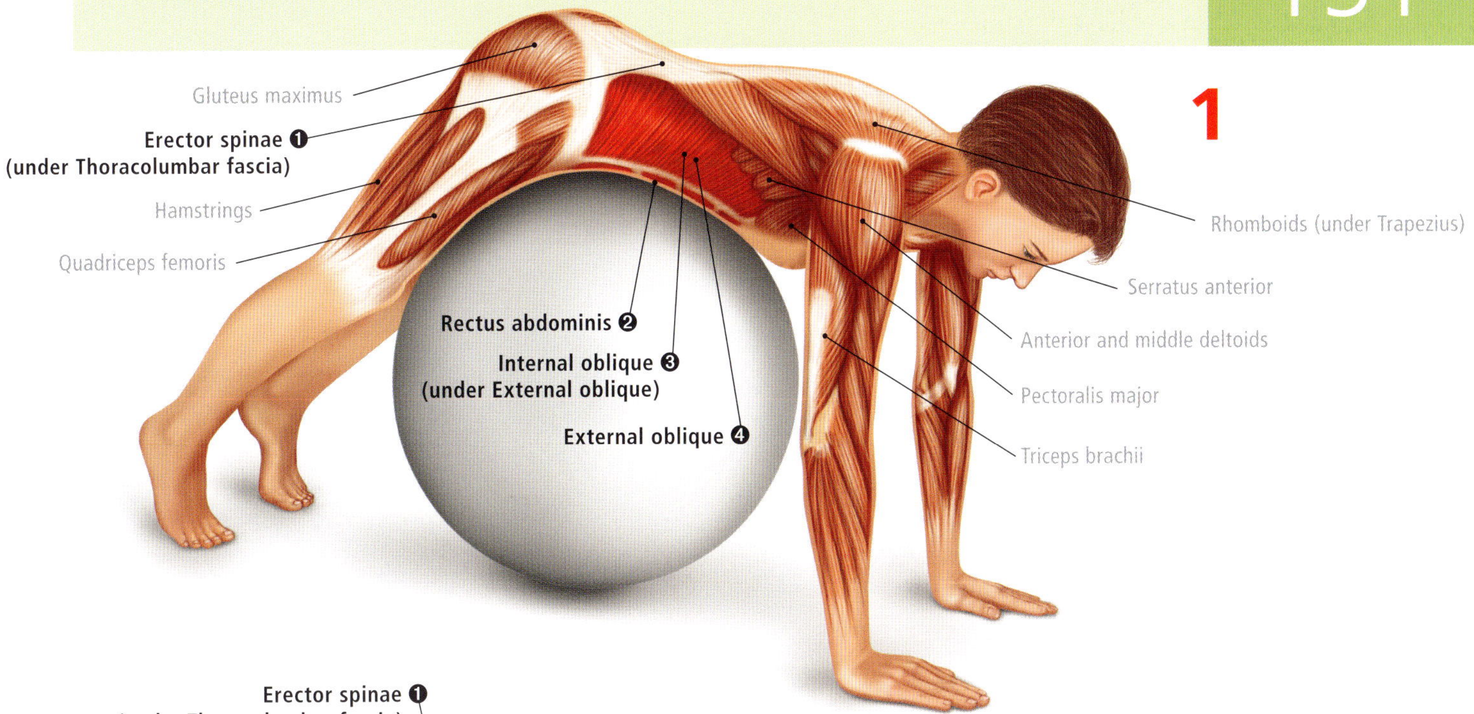

1

Gluteus maximus

Erector spinae ❶
(under Thoracolumbar fascia)

Hamstrings

Quadriceps femoris

Rectus abdominis ❷

Internal oblique ❸
(under External oblique)

External oblique ❹

Rhomboids (under Trapezius)

Serratus anterior

Anterior and middle deltoids

Pectoralis major

Triceps brachii

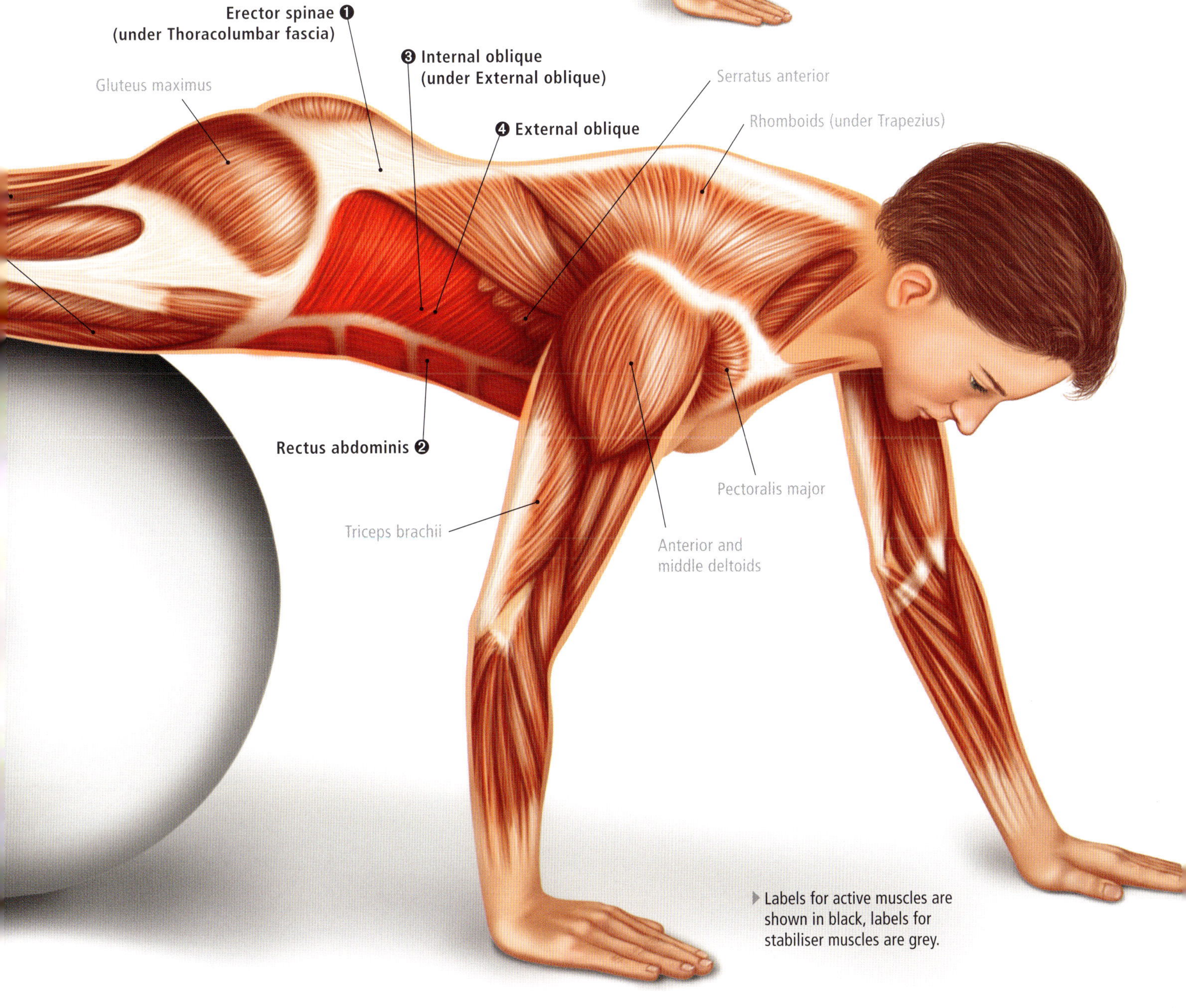

Erector spinae ❶
(under Thoracolumbar fascia)

❸ Internal oblique
(under External oblique)

❹ External oblique

Serratus anterior

Rhomboids (under Trapezius)

Gluteus maximus

Rectus abdominis ❷

Triceps brachii

Pectoralis major

Anterior and
middle deltoids

▶ Labels for active muscles are
shown in black, labels for
stabiliser muscles are grey.

Colouring Workbook

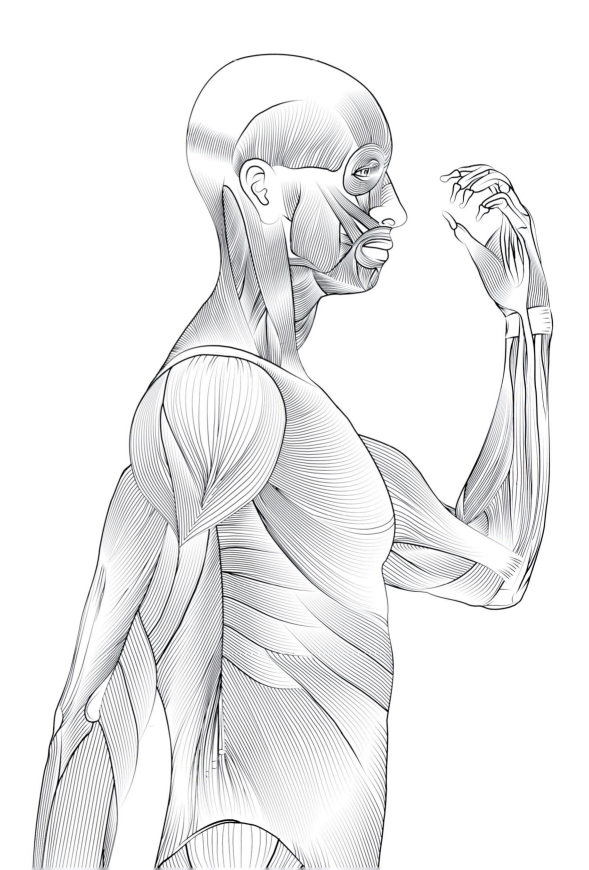

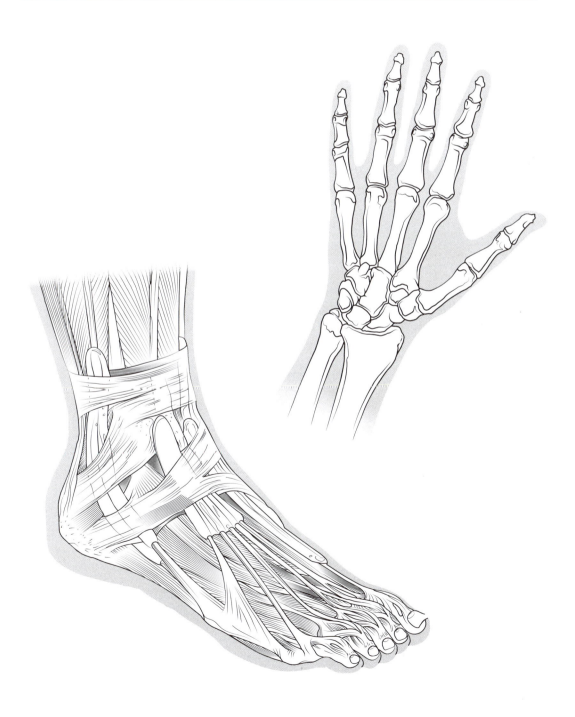

Muscular System

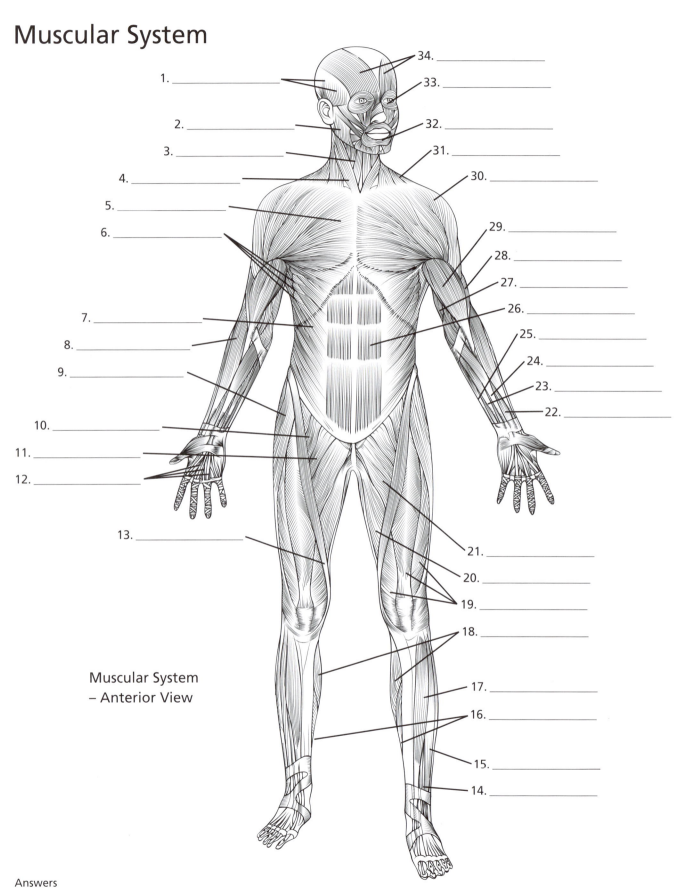

1. _____
2. _____
3. _____
4. _____
5. _____
6. _____
7. _____
8. _____
9. _____
10. _____
11. _____
12. _____
13. _____

34. _____
33. _____
32. _____
31. _____
30. _____
29. _____
28. _____
27. _____
26. _____
25. _____
24. _____
23. _____
22. _____
21. _____
20. _____
19. _____
18. _____
17. _____
16. _____
15. _____
14. _____

Muscular System
– Anterior View

Muscular System – Posterior View

Muscular System – Lateral View

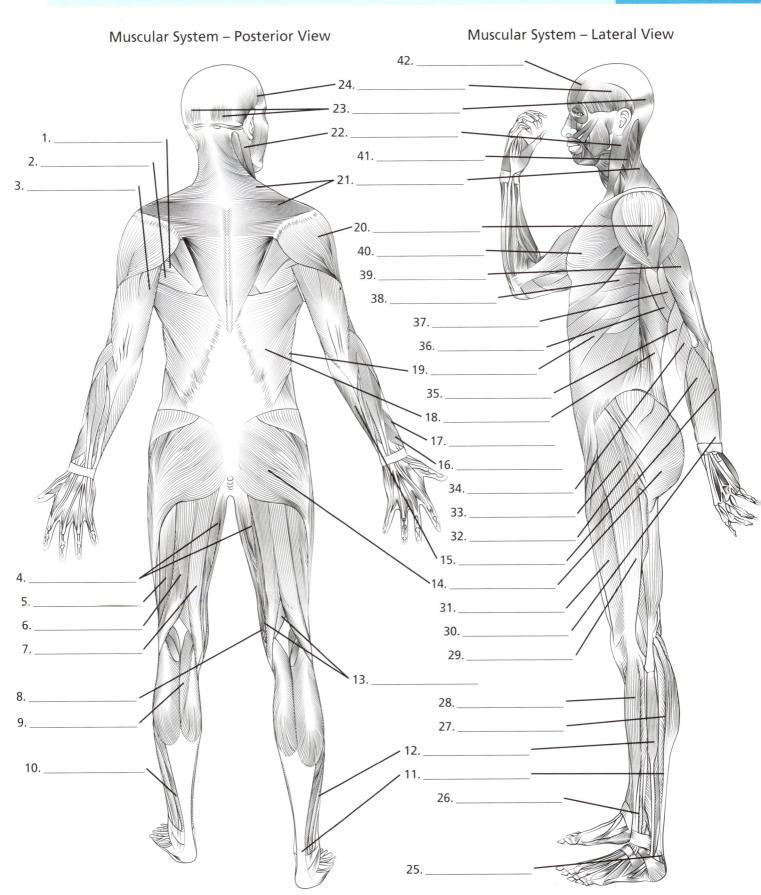

1. _____
2. _____
3. _____

4. _____
5. _____
6. _____
7. _____

8. _____
9. _____

10. _____

42. _____
24. _____
23. _____
22. _____
41. _____
21. _____

20. _____
40. _____
39. _____
38. _____

37. _____
36. _____
19. _____

35. _____
18. _____
17. _____
16. _____
34. _____
33. _____
32. _____
15. _____
14. _____
31. _____
30. _____
29. _____

13. _____

28. _____
27. _____

12. _____
11. _____
26. _____

25. _____

Answers

1. Teres minor, 2. Teres major, 3. Triceps brachii, 4. Adductor magnus, 5. Vastus lateralis, 6. Long head of biceps femoris, 7. Semitendinosus, 8. Gracilis, 9. Gastrocnemius, 10. Soleus, 11. Tendo calcaneus (Achilles tendon), 12. Fibularis (peroneus) longus, 13. Semimembranosus, 14. Gluteus maximus, 15. Flexor carpi ulnaris, 16. Extensor pollicis brevis, 17. Abductor pollicis longus, 18. Latissimus dorsi, 19. External abdominal oblique, 20. Deltoid, 21. Trapezius, 22. Sternocleidomastoid, 23. Occipitalis, 24. Temporalis, 25. Tendo calcaneus (Achilles tendon), 26. Extensor digitorum, 27. Lateral head of gastrocnemius, 28. Tibialis anterior, 29. Iliotibial tract, 30. Extensor carpi ulnaris, 31. Quadriceps femoris (vastus lateralis), 32. Extensor digitorum, 33. Tensor fasciae latae, 34. Extensor carpi radialis longus, 35. Brachioradialis, 36. Biceps brachii, 37. Brachialis, 38. Serratus anterior, 39. Lateral head of triceps brachii, 40. Pectoralis major, 41. Levator scapulae, 42. Frontalis

Muscles of the Head and Neck

Superficial and
Deep Muscles of
the Head and Neck –
Anterior View

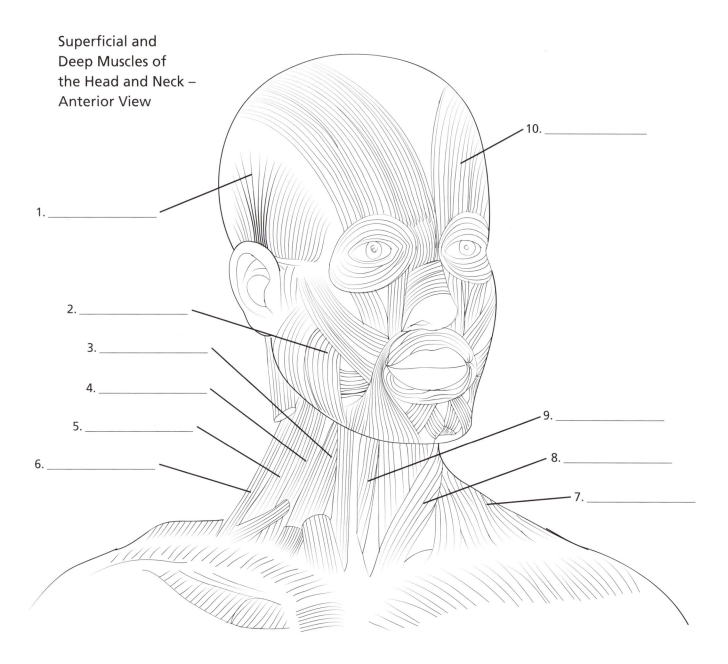

1. _____

2. _____

3. _____

4. _____

5. _____

6. _____

7. _____

8. _____

9. _____

10. _____

Superficial Muscles of the Head and Neck – Lateral View

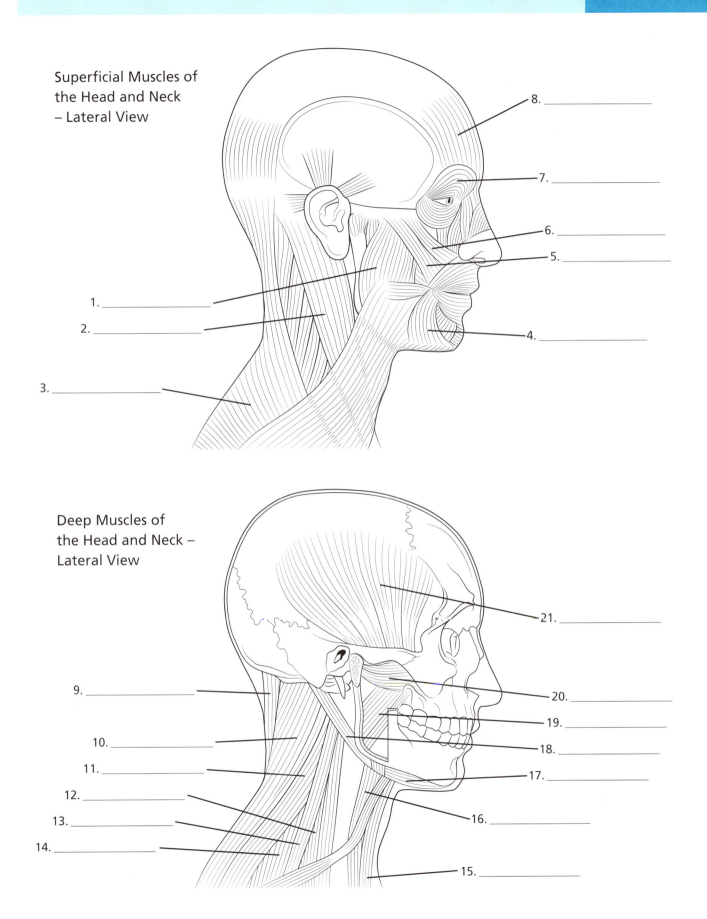

1. _____
2. _____
3. _____
4. _____
5. _____
6. _____
7. _____
8. _____

Deep Muscles of the Head and Neck – Lateral View

9. _____
10. _____
11. _____
12. _____
13. _____
14. _____
15. _____
16. _____
17. _____
18. _____
19. _____
20. _____
21. _____

Answers

1. Masseter, 2. Sternocleidomastoid, 3. Trapezius, 4. Depressor anguli oris, 5. Zygomaticus major, 6. Zygomaticus minor, 7. Orbicularis oculi, 8. Frontalis, 9. Semispinalis capitis, 10. Splenius capitis, 11. Levator scapulae, 12. Scalenus anterior, 13. Scalenus medius, 14. Scalenus posterior, 15. Sternohyoid, 16. Thyrohyoid, 17. Digastric (anterior belly), 18. Digastric (posterior belly), 19. Medial pterygoid, 20. Lateral pterygoid, 21. Temporalis

Muscles of the Back

Superficial Muscles
of the Back –
Posterior View

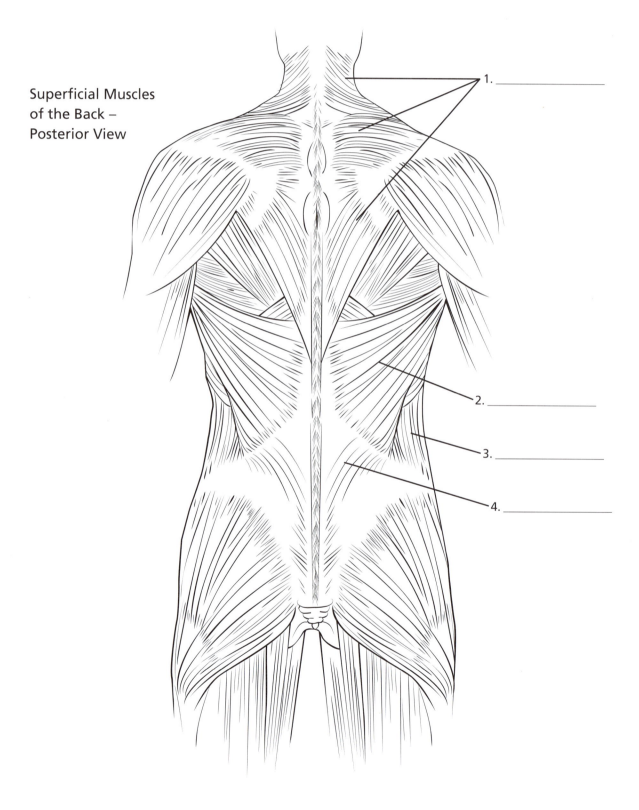

1. _____

2. _____

3. _____

4. _____

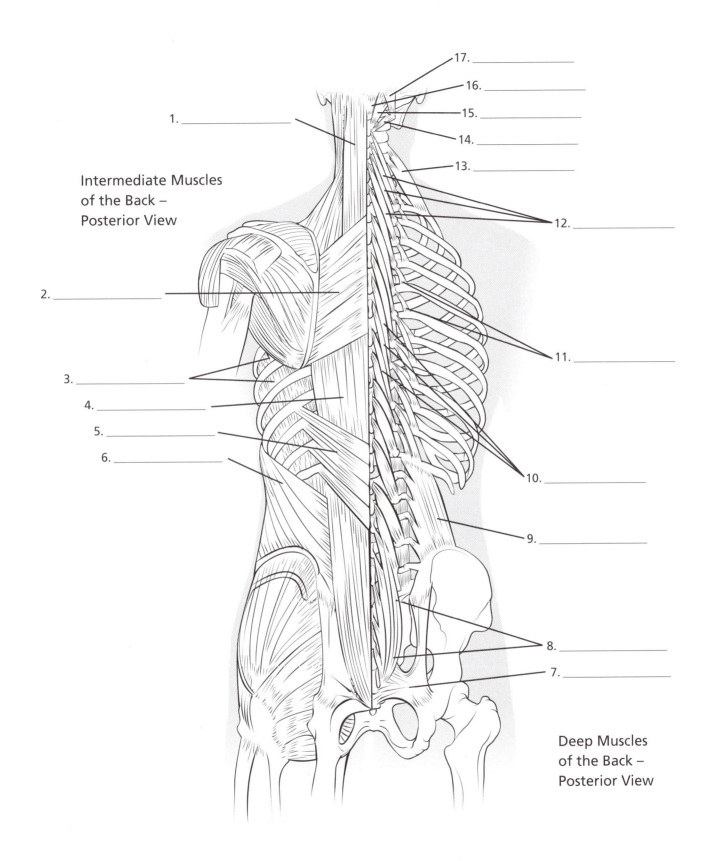

1. _____

**Intermediate Muscles
of the Back –
Posterior View**

2. _____

3. _____

4. _____

5. _____

6. _____

17. _____

16. _____

15. _____

14. _____

13. _____

12. _____

11. _____

10. _____

9. _____

8. _____

7. _____

**Deep Muscles
of the Back –
Posterior View**

Answers

1. Semispinalis capitis, 2. Rhomboid major, 3. External intercostals, 4. Erector spinae, 5. Serratus posterior inferior, 6. Internal oblique, 7. Sacrotuberous ligament, 8. Multifidus, 9. Quadratus lumborum, 10. Semispinalis thoracis, 11. Levatores costarum, 12. Semispinalis cervicis, 13. Scalenus posterior, 14. Obliquus capitis inferior, 15. Rectus capitis posterior major, 16. Rectus capitis posterior minor, 17. Obliquus capitis superior.

Muscles of the Thorax and Abdomen

Superficial and Deep Muscles
of the Thorax and Abdomen
– Anterior View

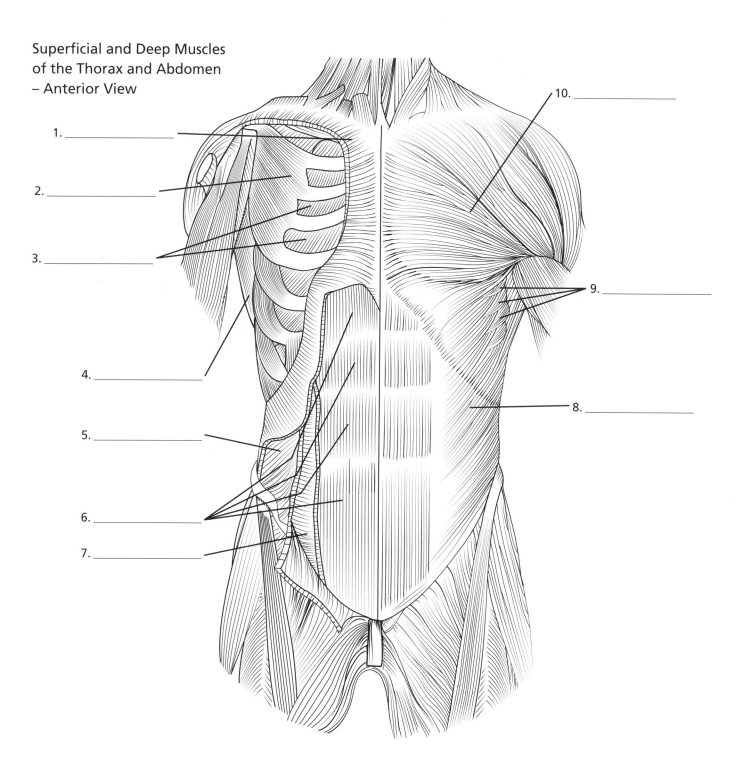

1. _____

2. _____

3. _____

4. _____

5. _____

6. _____

7. _____

8. _____

9. _____

10. _____

Answers

Muscles of the Shoulder

Superficial and Deep
Muscles of the Shoulder
– Posterior View

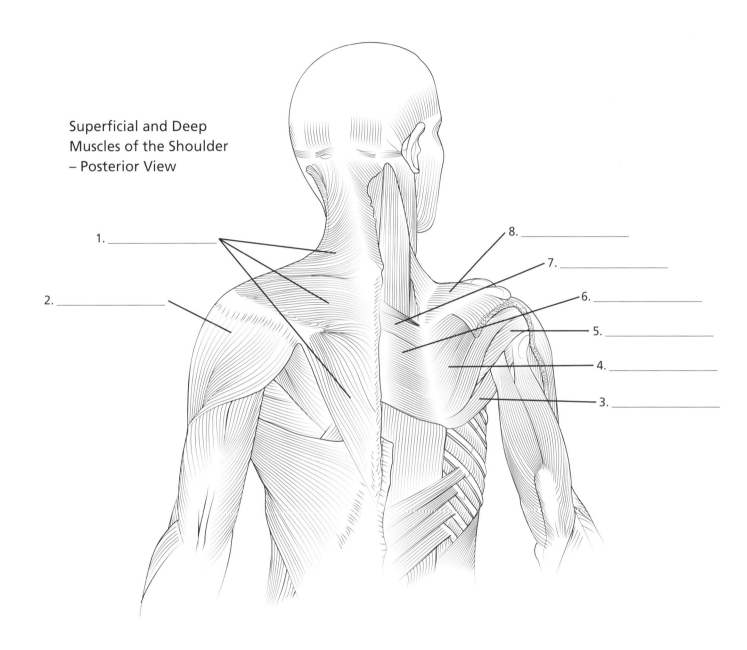

1. _____

2. _____

8. _____

7. _____

6. _____

5. _____

4. _____

3. _____

Answers

Muscles of the Shoulder

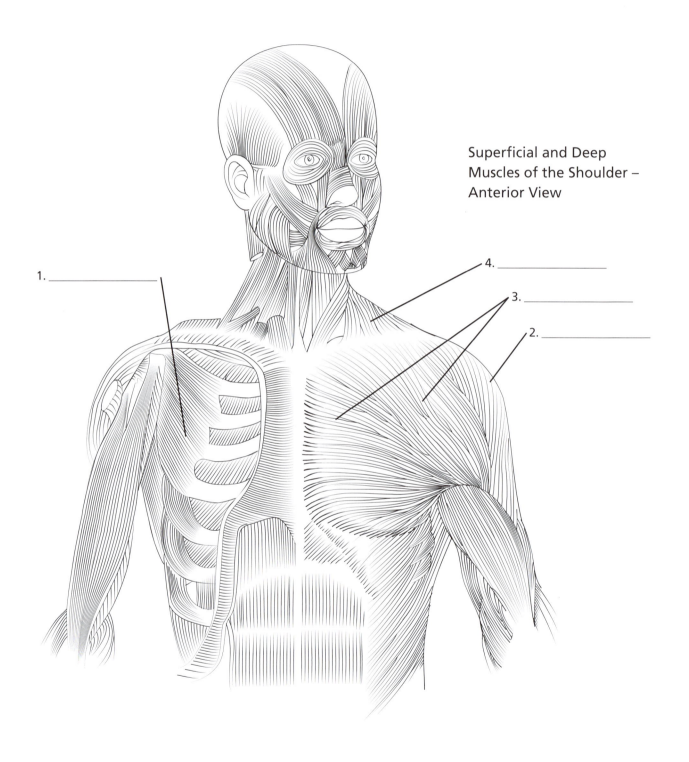

Superficial and Deep
Muscles of the Shoulder –
Anterior View

1. _____

4. _____

3. _____

2. _____

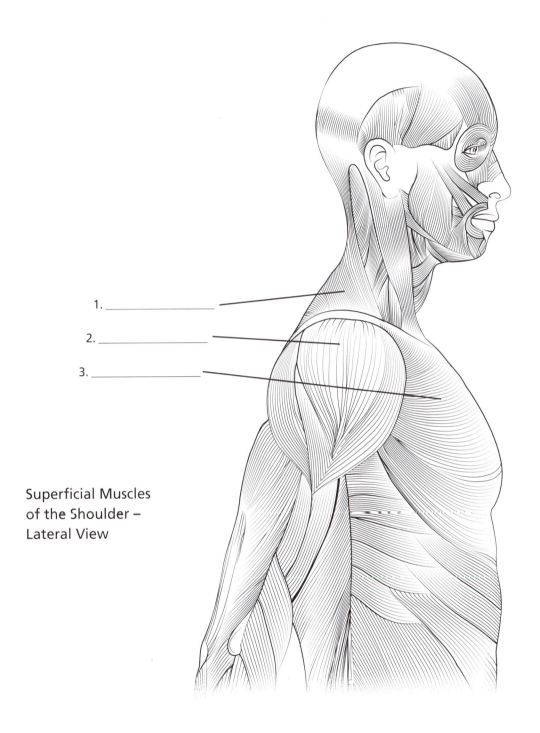

1. _____

2. _____

3. _____

Superficial Muscles of the Shoulder – Lateral View

Answers

1. Trapezius, 2. Deltoid, 3. Pectoralis major

Muscles of the Upper Limb

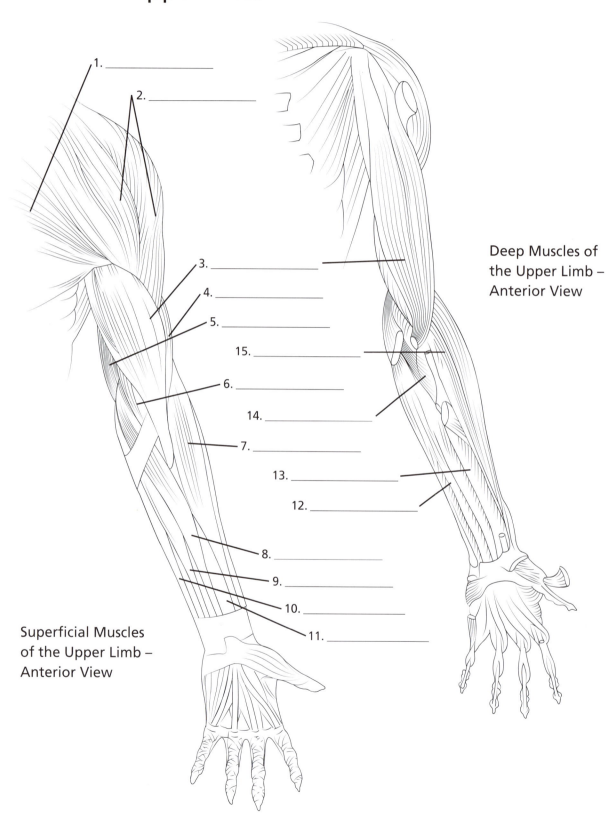

1. _____

2. _____

Deep Muscles of
the Upper Limb –
Anterior View

3. _____

4. _____

5. _____

15. _____

6. _____

14. _____

7. _____

13. _____

12. _____

8. _____

9. _____

10. _____

11. _____

Superficial Muscles
of the Upper Limb –
Anterior View

Answers

1. Pectoralis major, 2. Deltoid, 3. Biceps brachii, 4. Brachialis, 5. Triceps brachii, 6. Pronator teres, 7. Brachioradialis, 8. Tendon of flexor carpi radialis, 9. Tendon of palmaris longus, 10. Tendon of flexor carpi ulnaris, 11. Flexor digitorum superficialis, 12. Flexor digitorum profundus, 13. Flexor pollicis longus, 14. Pronator teres, 15. Extensor carpi radialis longus

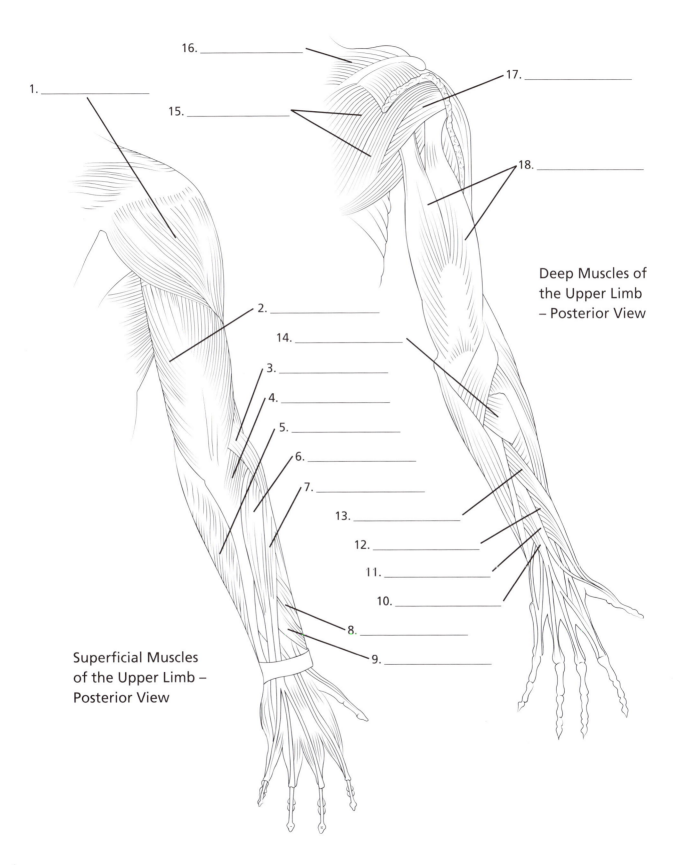

16. _____

1. _____

15. _____

17. _____

18. _____

Deep Muscles of
the Upper Limb
– Posterior View

2. _____

14. _____

3. _____

4. _____

5. _____

6. _____

7. _____

13. _____

12. _____

11. _____

10. _____

8. _____

9. _____

Superficial Muscles
of the Upper Limb –
Posterior View

Muscles of the Upper Limb

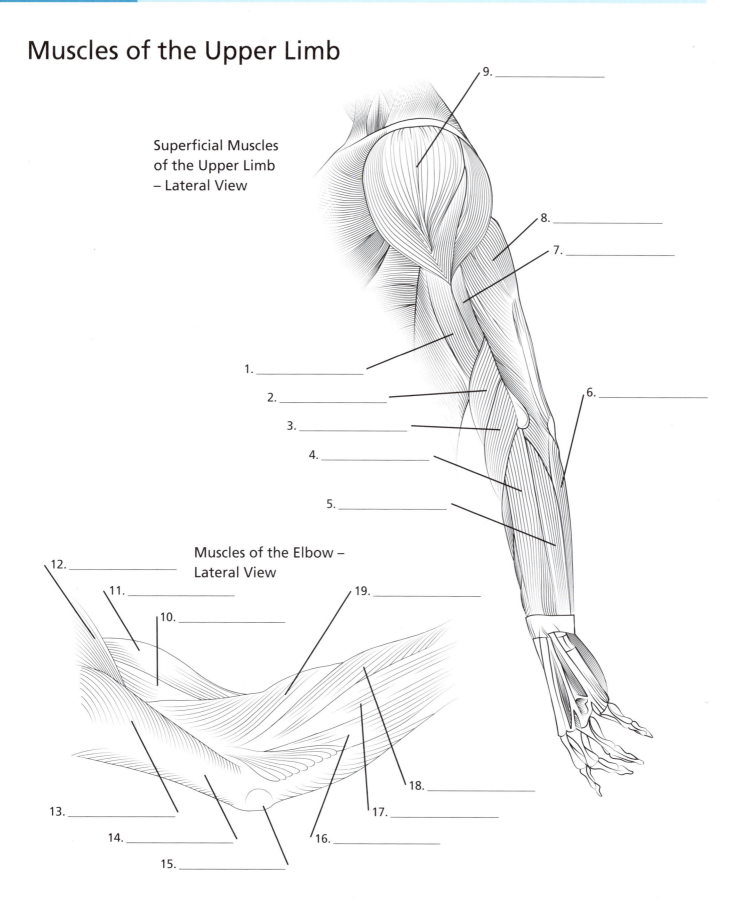

Superficial Muscles
of the Upper Limb
– Lateral View

9. _____

8. _____

7. _____

1. _____

2. _____

3. _____

4. _____

5. _____

6. _____

Muscles of the Elbow –
Lateral View

12. _____

11. _____

10. _____

19. _____

13. _____

14. _____

15. _____

16. _____

17. _____

18. _____

Answers

1. Biceps brachii, 2. Brachioradialis, 3. Extensor carpi radialis longus, 4. Extensor digitorum, 5. Extensor carpi ulnaris, 6. Flexor carpi ulnaris, 7. Brachialis, 8. Lateral head of triceps brachii, 9. Deltoid, 10. Brachialis, 11. Biceps brachii, 12. Deltoid, 13. Triceps brachii, 14. Tendon of triceps brachii, 15. Olecranon, 16. Extensor carpi ulnaris, 17. Extensor digitorum, 18. Extensor carpi radialis longus, 19. Brachioradialis

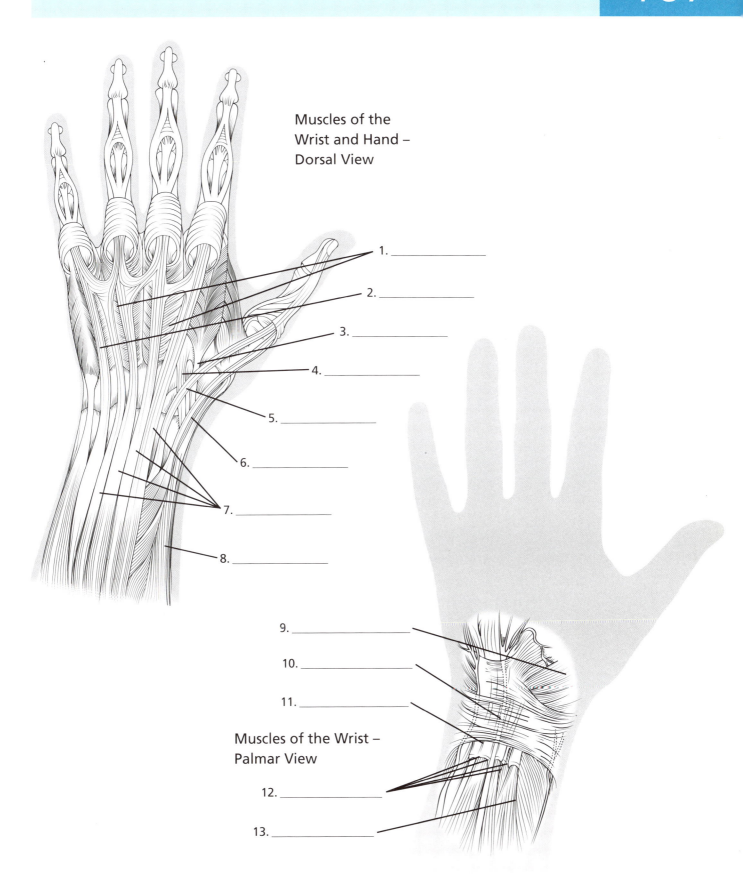

Muscles of the
Wrist and Hand –
Dorsal View

1. _____

2. _____

3. _____

4. _____

5. _____

6. _____

7. _____

8. _____

9. _____

10. _____

11. _____

Muscles of the Wrist –
Palmar View

12. _____

13. _____

Answers

Muscles of the Lower Limb

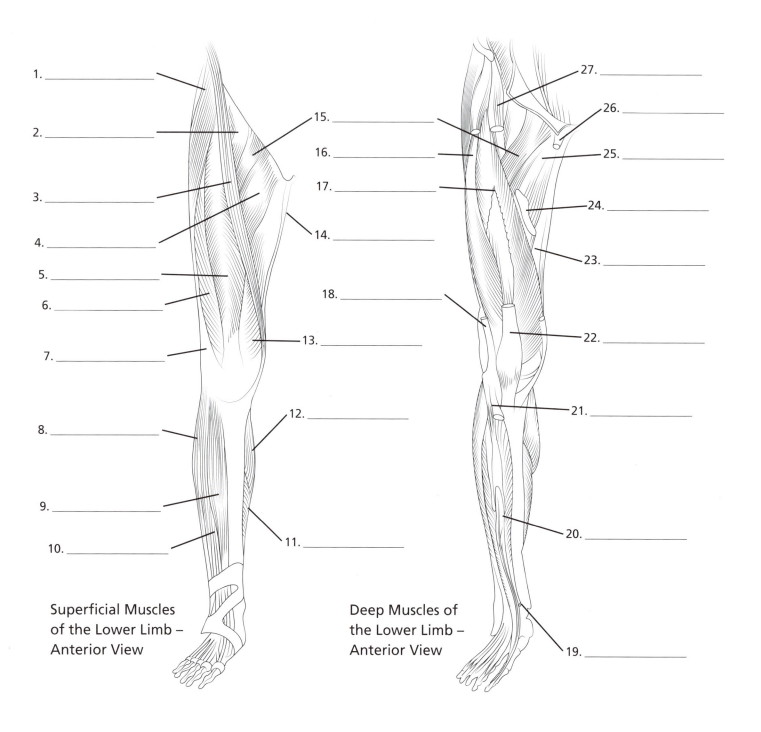

1. _____

2. _____

3. _____

4. _____

5. _____

6. _____

7. _____

8. _____

9. _____

10. _____

15. _____

16. _____

17. _____

14. _____

18. _____

13. _____

12. _____

11. _____

27. _____

26. _____

25. _____

24. _____

23. _____

22. _____

21. _____

20. _____

19. _____

Superficial Muscles
of the Lower Limb –
Anterior View

Deep Muscles of
the Lower Limb –
Anterior View

Answers

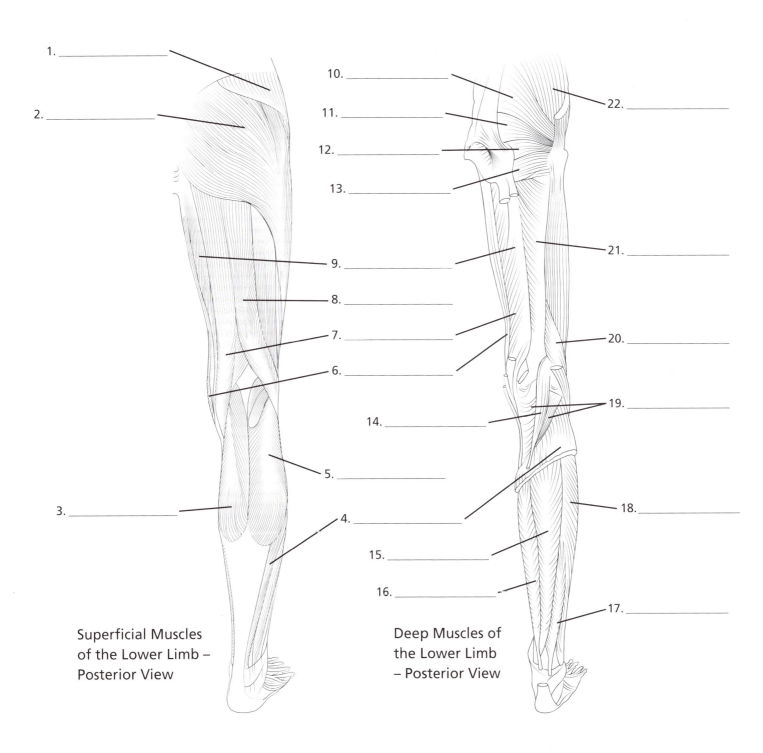

1. _____

2. _____

3. _____

10. _____

11. _____

12. _____

13. _____

9. _____

8. _____

7. _____

6. _____

14. _____

4. _____

15. _____

16. _____

5. _____

22. _____

21. _____

20. _____

19. _____

18. _____

17. _____

Superficial Muscles
of the Lower Limb –
Posterior View

Deep Muscles of
the Lower Limb
– Posterior View

Answers

1. Gluteus medius, 2. Gluteus maximus, 3. Medial head of gastrocnemius, 4. Soleus, 5. Lateral head of gastrocnemius, 6. Gracilis, 7. Semitendinosus, 8. Biceps femoris, 9. Adductor magnus, 10. Piriformis, 11. Superior gemellus, 12. Inferior gemellus, 13. Quadratus femoris, 14. Plantaris, 15. Tibialis posterior, 16. Flexor digitorum longus, 17. Flexor hallucis longus, 18. Fibularis (peroneus) longus, 19. Popliteus, 20. Short head of biceps femoris, 21. Adductor part of adductor magnus, 22. Gluteus minimus

Muscles of the Lower Limb

**Superficial Muscles
of the Lower Limb
– Lateral View**

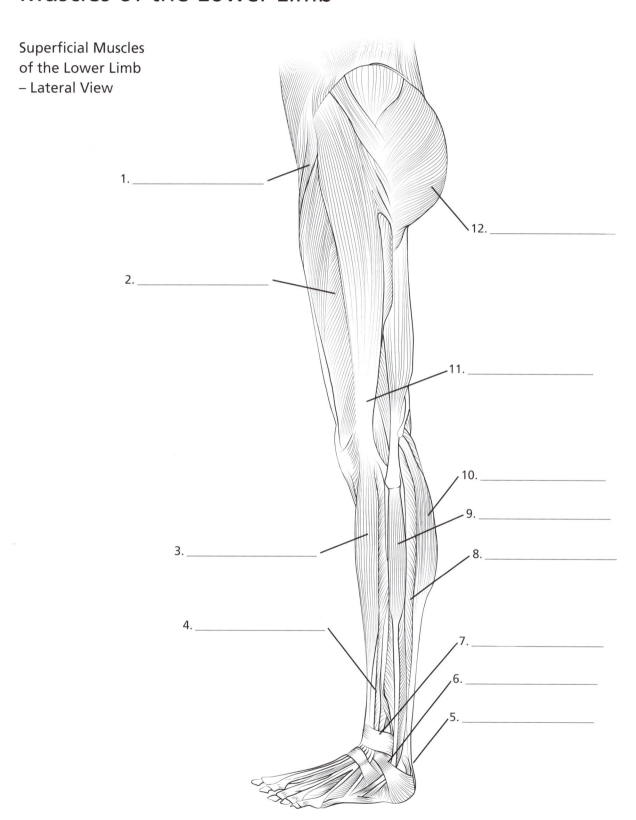

1. _____

2. _____

3. _____

4. _____

5. _____

6. _____

7. _____

8. _____

9. _____

10. _____

11. _____

12. _____

Answers

1. Sartorius, 2. Quadriceps femoris (vastus lateralis), 3. Tibialis anterior, 4. Extensor digitorum longus, 5. Tendo calcaneus (Achilles tendon), 6. Inferior extensor retinaculum, 7. Superior extensor retinaculum, 8. Soleus, 9. Fibularis (peroneus) longus, 10. Lateral head of gastrocnemius, 11. Illiotibial tract, 12. Gluteus maximus

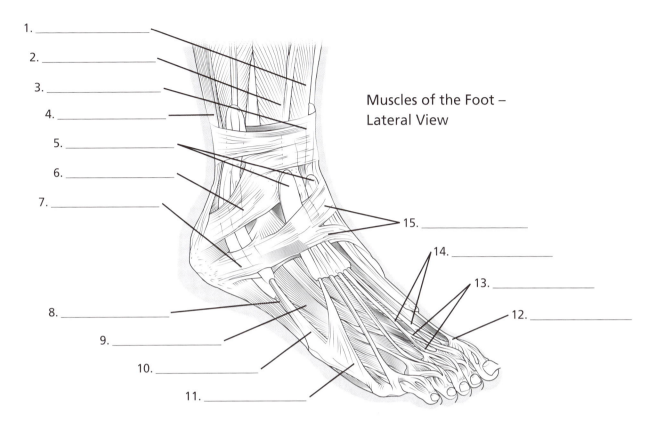

Muscles of the Foot –
Lateral View

1. _____
2. _____
3. _____
4. _____
5. _____
6. _____
7. _____
8. _____
9. _____
10. _____
11. _____
12. _____
13. _____
14. _____
15. _____

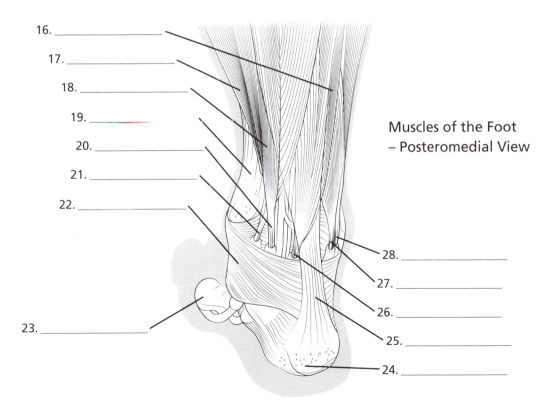

Muscles of the Foot
– Posteromedial View

16. _____
17. _____
18. _____
19. _____
20. _____
21. _____
22. _____
23. _____
24. _____
25. _____
26. _____
27. _____
28. _____

Answers

Muscle Types

1. _____

2. _____

3. _____

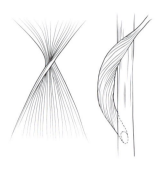

4. _____

5. _____

6. _____

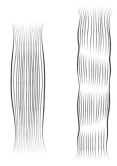

7. _____

8. _____

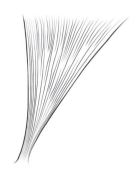

9. _____

10. _____

11. _____

12. _____

13. _____

14. _____

15. _____

16. _____

Answers

Articulations

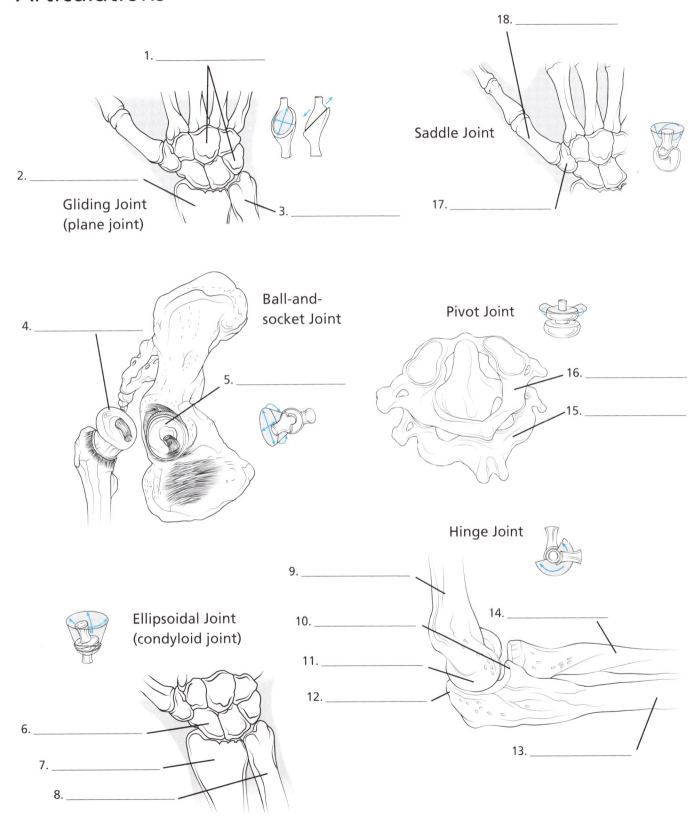

1. _____

2. _____

**Gliding Joint
(plane joint)**

3. _____

18. _____

Saddle Joint

17. _____

4. _____

**Ball-and-
socket Joint**

5. _____

Pivot Joint

16. _____

15. _____

**Ellipsoidal Joint
(condyloid joint)**

6. _____

7. _____

8. _____

Hinge Joint

9. _____

10. _____

11. _____

12. _____

14. _____

13. _____

Answers

1. Carpal bones, 2. Radius, 3. Ulna, 4. Head of femur (ball), 5. Acetabulum (socket), 6. Scaphoid bone, 7. Radius, 8. Ulna, 9. Humerus, 10. Coronoid process of ulna, 11. Trochlea (of humerus), 12. Olecranon, 13. Ulna, 14. Radius, 15. Axis, 16. Atlas, 17. Trapezium bone, 18. Metacarpal bone of thumb

Skeletal System

Skeletal System – Anterior View

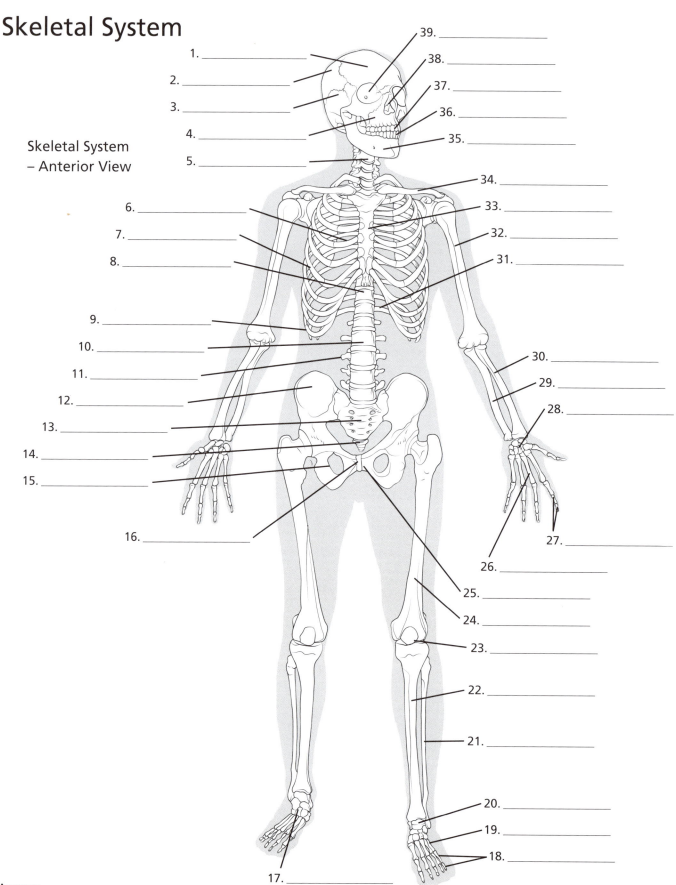

1. _____
2. _____
3. _____
4. _____
5. _____
6. _____
7. _____
8. _____
9. _____
10. _____
11. _____
12. _____
13. _____
14. _____
15. _____
16. _____
17. _____
18. _____
19. _____
20. _____
21. _____
22. _____
23. _____
24. _____
25. _____
26. _____
27. _____
28. _____
29. _____
30. _____
31. _____
32. _____
33. _____
34. _____
35. _____
36. _____
37. _____
38. _____
39. _____

Answers

1. Frontal bone, 2. Parietal bone, 3. Temporal bone, 4. Maxilla, 5. Cervical vertebra, 6. Costal cartilage, 7. True rib, 8. Thoracic vertebra, 9. False rib, 10. Lumbar vertebra, 11. Transverse process, 12. Ilium, 13. Sacrum, 14. Coccyx, 15. Ischium, 16. Pubic symphysis, 17. Tarsal bones, 18. Phalanges, 19. Metatarsal bones, 20. Talus, 21. Fibula, 22. Tibia, 23. Patella, 24. Femur, 25. Pubis, 26. Metacarpal bones, 27. Phalanges, 28. Carpal bones, 29. Ulna, 30. Radius, 31. Twelfth rib (floating rib), 32. Humerus, 33. Sternum, 34. Clavicle, 35. Mandible, 36. Lower teeth, 37. Upper teeth, 38. Anterior nasal (piriform) aperture, 39. Orbit

Skeletal System – Posterior View

Skeletal System – Lateral View

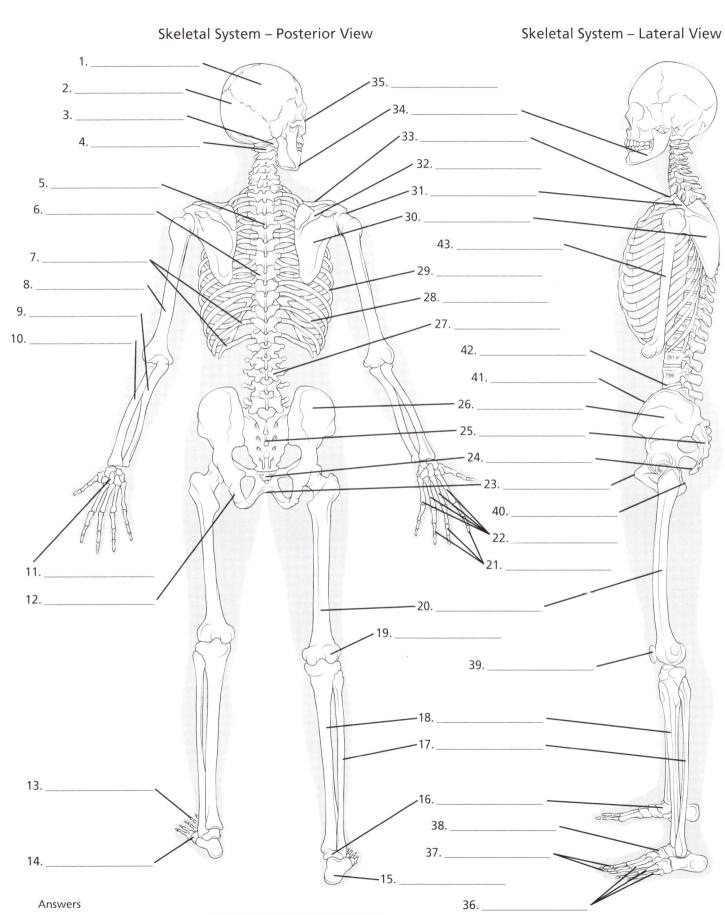

1. _____
2. _____
3. _____
4. _____
5. _____
6. _____
7. _____
8. _____
9. _____
10. _____
11. _____
12. _____
13. _____
14. _____

35. _____
34. _____
33. _____
32. _____
31. _____
30. _____
43. _____
29. _____
28. _____
27. _____
42. _____
41. _____
26. _____
25. _____
24. _____
23. _____
40. _____
22. _____
21. _____
20. _____
19. _____
39. _____
18. _____
17. _____
16. _____
38. _____
37. _____
15. _____
36. _____

Answers

Vertebral Column

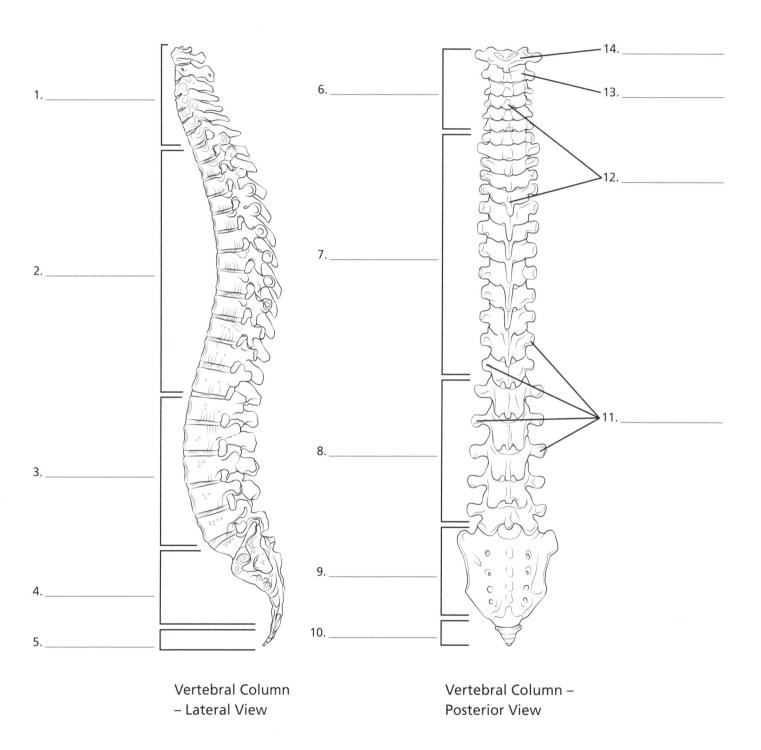

1. _____

2. _____

3. _____

4. _____

5. _____

6. _____

7. _____

8. _____

9. _____

10. _____

11. _____

12. _____

13. _____

14. _____

Vertebral Column
– Lateral View

Vertebral Column –
Posterior View

Answers

1. Cervical vertebrae (C1–C7), 2. Thoracic vertebrae (T1–T12), 3. Lumbar vertebrae (L1–L5), 4. Sacrum, 5. Coccyx, 6. Cervical region (C1–C7), 7. Thoracic region (T1–T12), 8. Lumbar region (L1–L5), 9. Sacral region (S1–S5), 10. Coccygeal region, 11. Transverse processes, 12. Spinous processes, 13. Axis (C2), 14. Atlas (C1)

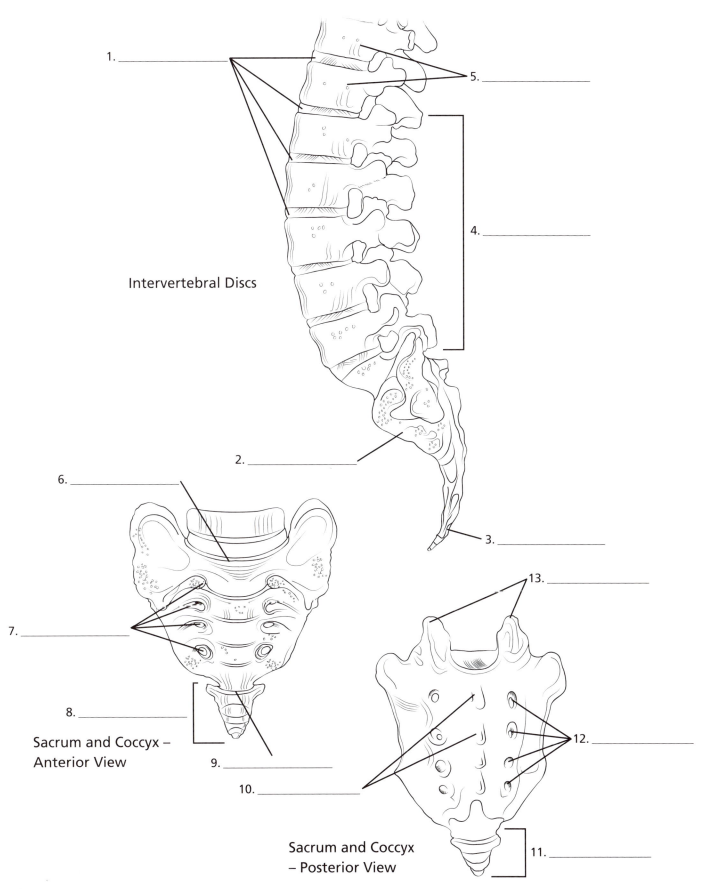

Intervertebral Discs

1. _____

5. _____

4. _____

2. _____

6. _____

3. _____

7. _____

8. _____

9. _____

13. _____

12. _____

10. _____

11. _____

Sacrum and Coccyx –
Anterior View

Sacrum and Coccyx
– Posterior View

Answers

1. Intervertebral discs, 2. Sacrum, 3. Coccyx, 4. Lumbar vertebrae, 5. Thoracic vertebrae, 6. Sacral promontory, 7. Pelvic sacral foramina, 8. Coccyx, 9. Sacrococcygeal joint, 10. Median sacral crest and spinous tubercles, 11. Coccyx, 12. Posterior sacral foramina, 13. Superior articular processes (facets)

Bones of the Upper Limb

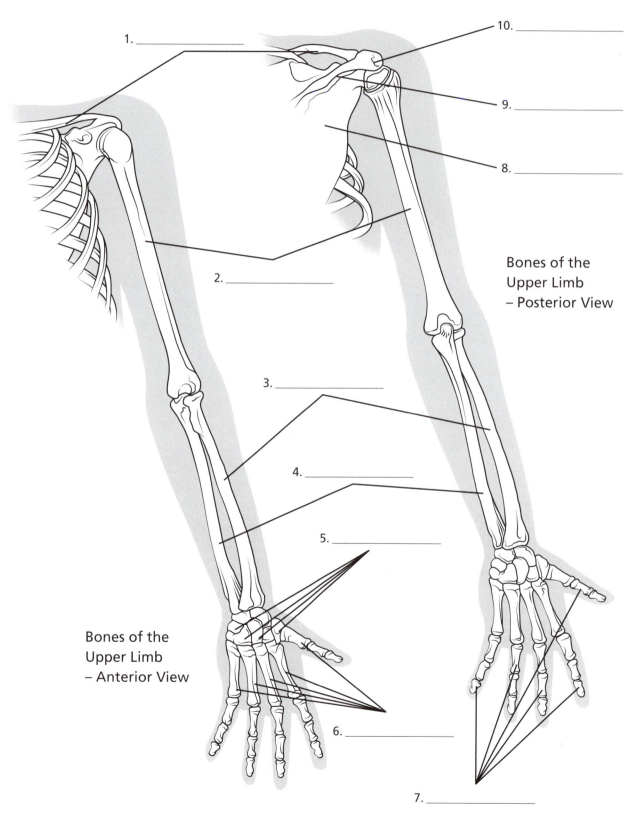

1. _____

10. _____

9. _____

8. _____

Bones of the
Upper Limb
– Posterior View

2. _____

3. _____

4. _____

5. _____

Bones of the
Upper Limb
– Anterior View

6. _____

7. _____

Answers

1. Clavicle, 2. Humerus, 3. Radius, 4. Ulna, 5. Carpal bones, 6. Metacarpal bones, 7. Phalanges, 8. Scapula, 9. Spine of scapula, 10. Acromion

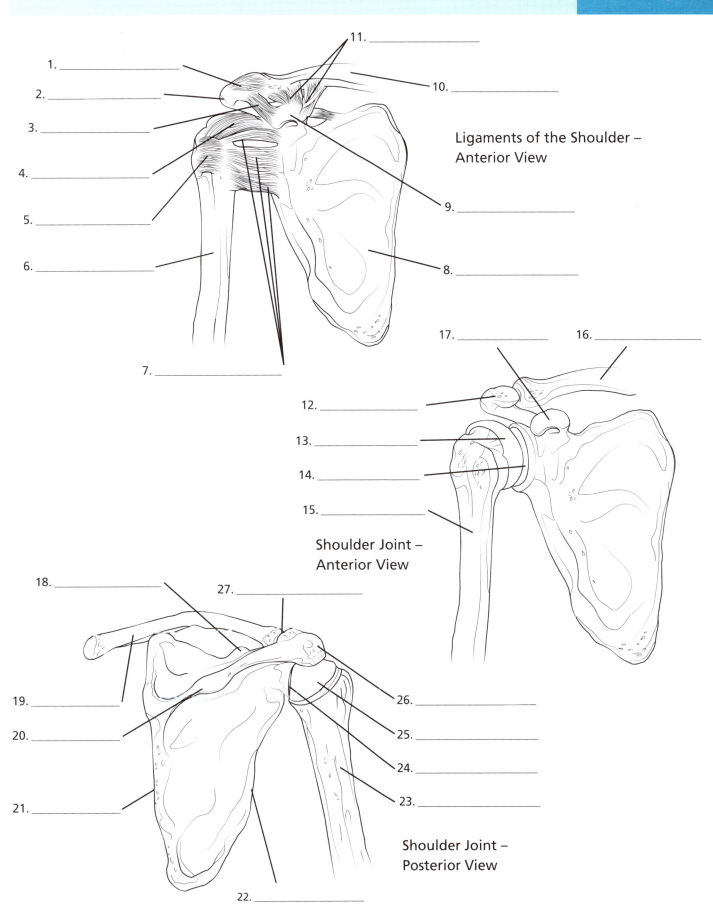

1. _____
2. _____
3. _____
4. _____
5. _____
6. _____
7. _____

11. _____
10. _____

**Ligaments of the Shoulder –
Anterior View**

9. _____
8. _____

17. _____   16. _____

12. _____
13. _____
14. _____
15. _____

**Shoulder Joint –
Anterior View**

18. _____   27. _____

19. _____
20. _____
21. _____

26. _____
25. _____
24. _____
23. _____

22. _____

**Shoulder Joint –
Posterior View**

Answers

1. Acromioclavicular ligament, 2. Acromion, 3. Coracoacromial ligament, 4. Coracohumeral ligament, 5. Transverse humeral ligament, 6. Shaft of humerus, 7. Glenohumeral ligaments, 8. Scapula, 9. Coracoid process, 10. Clavicle, 11. Coracoclavicular ligament, 12. Acromion, 13. Head of humerus, 14. Glenoid cavity, 15. Humerus, 16. Clavicle, 17. Coracoid, 18. Coracoid process, 19. Clavicle, 20. Spine of scapula, 21. Medial border of scapula, 22. Lateral border of scapula, 23. Humerus, 24. Glenoid fossa, 25. Head of humerus, 26. Acromion, 27. Acromioclavicular joint

Bones of the Upper Limb

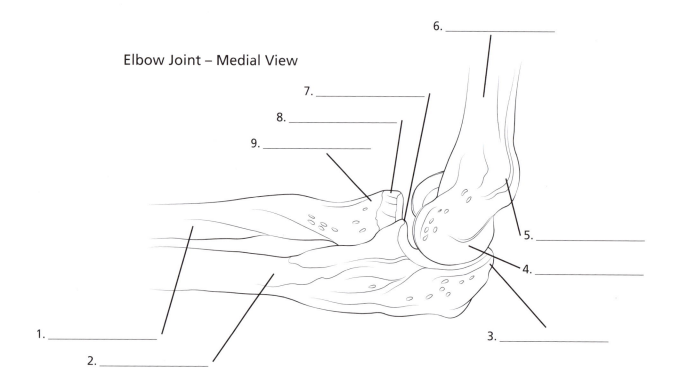

Elbow Joint – Medial View

6. _____

7. _____

8. _____

9. _____

5. _____

4. _____

1. _____

2. _____

3. _____

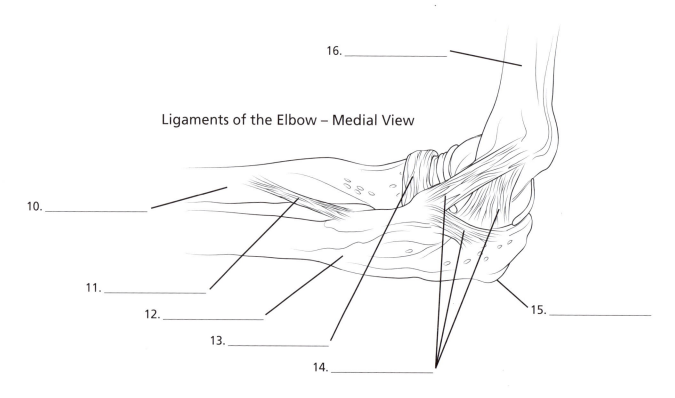

16. _____

Ligaments of the Elbow – Medial View

10. _____

11. _____

12. _____

13. _____

14. _____

15. _____

1. _____

2. _____

3. _____

**Bones of the
Wrist and Hand
– Dorsal View**

4. _____

5. _____

6. _____

7. _____

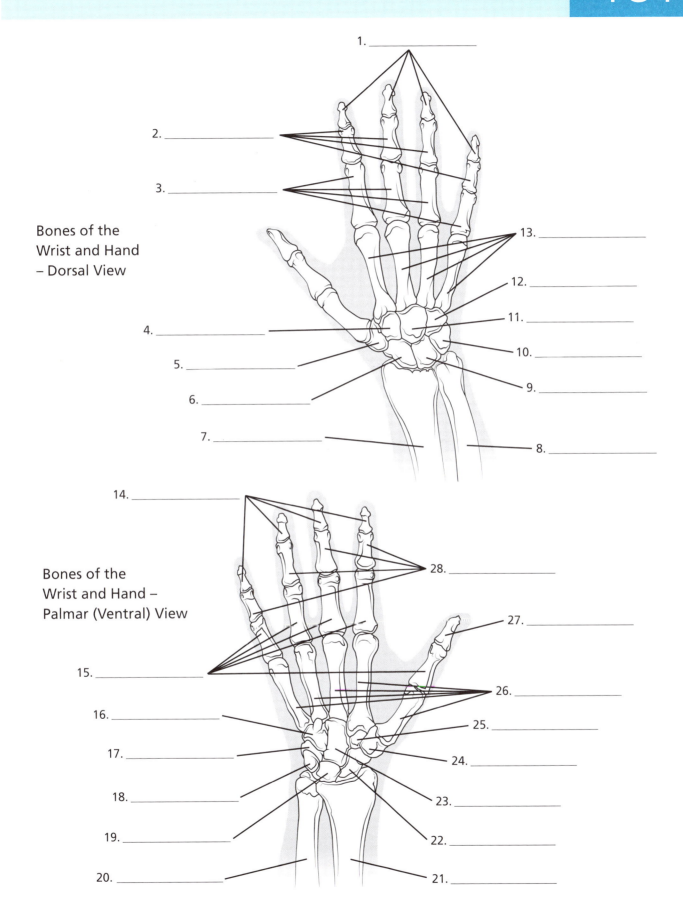

13. _____

12. _____

11. _____

10. _____

9. _____

8. _____

14. _____

**Bones of the
Wrist and Hand –
Palmar (Ventral) View**

15. _____

16. _____

17. _____

18. _____

19. _____

20. _____

28. _____

27. _____

26. _____

25. _____

24. _____

23. _____

22. _____

21. _____

Answers

1. Distal phalanges, 2. Middle phalanges, 3. Proximal phalanges, 4. Trapezoid, 5. Trapezium, 6. Scaphoid, 7. Radius, 8. Ulna, 9. Ulna, 10. Triquetrum, 11. Capitate, 12. Hamate, 13. Metacarpal bones, 14. Distal phalanges, 15. Proximal phalanges, 16. Hamate, 17. Triquetrum, 18. Pisiform, 19. Lunate, 20. Ulna, 21. Radius, 22. Scaphoid, 23. Capitate, 24. Trapezium, 25. Trapezoid, 26. Metacarpal bones, 27. Distal phalanx of thumb, 28. Middle phalanges

Bones of the Lower Limb

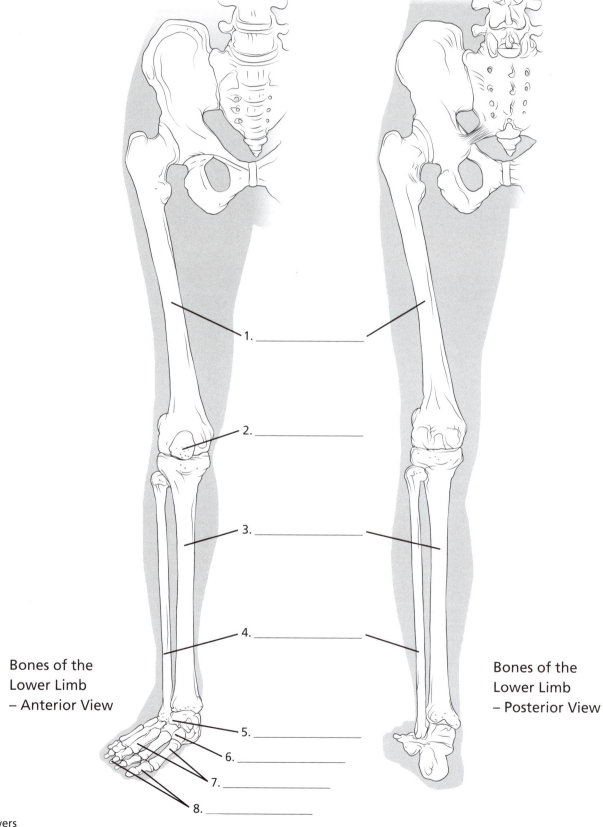

1. _____

2. _____

3. _____

4. _____

Bones of the
Lower Limb
– Anterior View

5. _____

6. _____

7. _____

8. _____

Bones of the
Lower Limb
– Posterior View

Answers

1. Femur, 2. Patella, 3. Tibia, 4. Fibula, 5. Talus, 6. Tarsal bones, 7. Metatarsal bones, 8. Phalanges

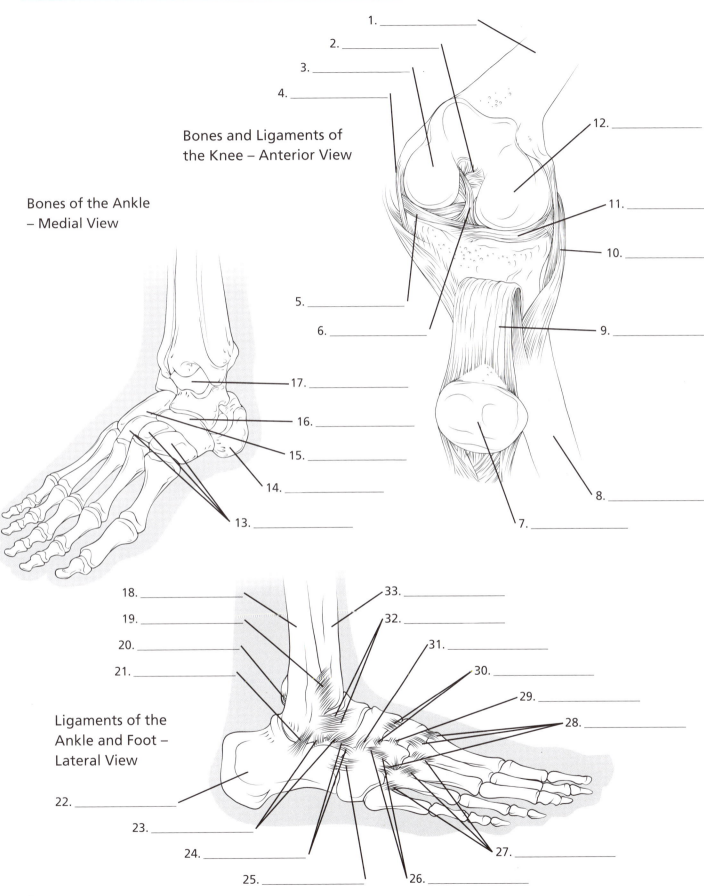

Bones and Ligaments of
the Knee – Anterior View

Bones of the Ankle
– Medial View

Ligaments of the
Ankle and Foot –
Lateral View

1. _____

2. _____

3. _____

4. _____

12. _____

11. _____

10. _____

9. _____

5. _____

6. _____

17. _____

16. _____

15. _____

14. _____

13. _____

8. _____

7. _____

18. _____

19. _____

20. _____

21. _____

22. _____

23. _____

24. _____

25. _____

33. _____

32. _____

31. _____

30. _____

29. _____

28. _____

27. _____

26. _____

Answers

1. Femur, 2. Posterior cruciate ligament, 3. Patellar ligament, 4. Fibular (lateral) collateral ligament, 5. Lateral meniscus, 6. Anterior cruciate ligament, 7. Patella (reflected), 8. Tibia, 9. Patellar ligament, 10. Tibial (medial) collateral ligament, 11. Medial meniscus, 12. Medial condyle of femur, 13. Cuneiform bones, 14. Calcaneus, 15. Cuboid, 16. Navicular, 17. Talus, 18. Fibula, 19. Anterior tibiofibular ligament, 20. Posterior tibiofibular ligament, 21. Calcaneofibular ligament, 22. Calcaneus, 23. Talocalcaneal ligaments, 24. Bifurcate ligament, 25. Dorsal calcaneocuboid ligament, 26. Dorsal cuneocuboid ligament, 27. Dorsal tarsometatarsal ligaments, 28. Dorsal metatarsal ligaments, 29. Dorsal intercuneiform ligament, 30. Dorsal cuneonavicular ligaments, 31. Dorsal cuboideonavicular ligament, 32. Anterior talofibular ligament, 33. Tibia

Nerves of the Upper and Lower Limb

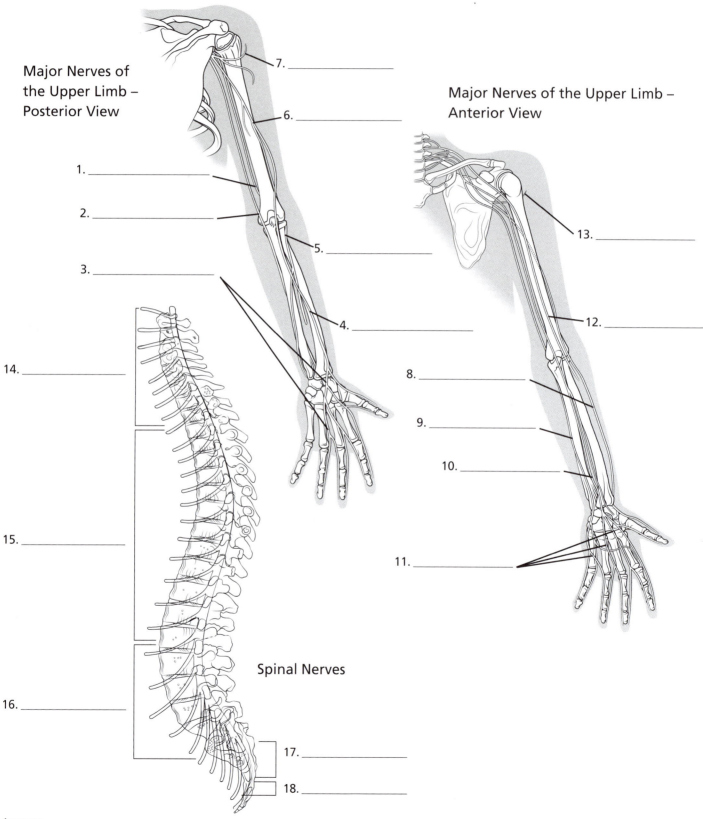

Major Nerves of the Upper Limb – Posterior View

7. _____

6. _____

1. _____

2. _____

3. _____

5. _____

4. _____

14. _____

15. _____

16. _____

Major Nerves of the Upper Limb – Anterior View

13. _____

12. _____

8. _____

9. _____

10. _____

11. _____

Spinal Nerves

17. _____

18. _____

Answers

Major Nerves of the Lower Limb – Posterior View

Major Nerves of the Lower Limb – Anterior View

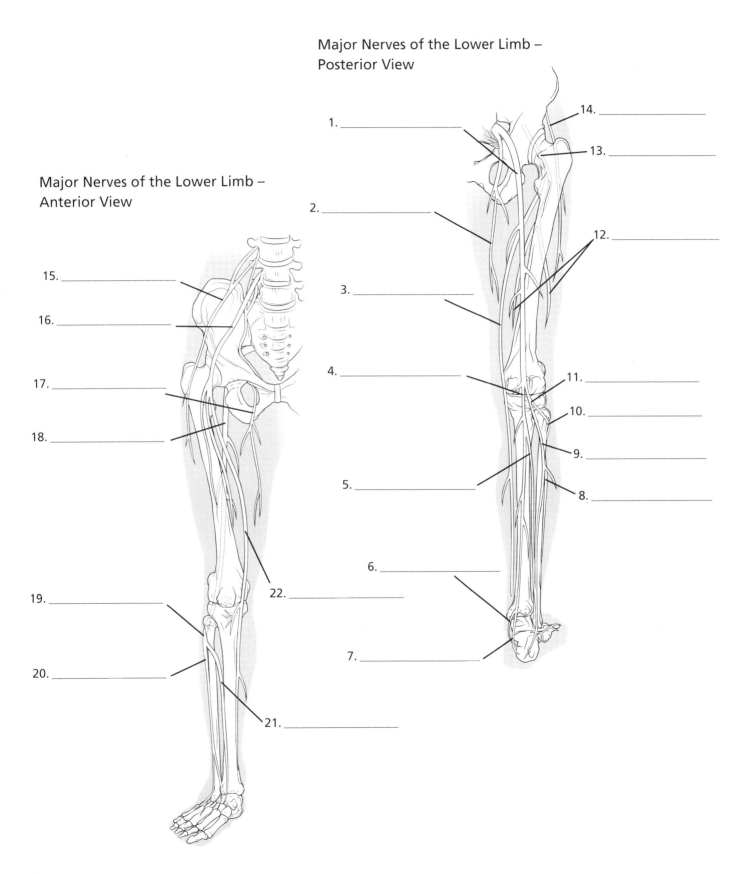

1. _____
2. _____
3. _____
4. _____
5. _____
6. _____
7. _____
8. _____
9. _____
10. _____
11. _____
12. _____
13. _____
14. _____
15. _____
16. _____
17. _____
18. _____
19. _____
20. _____
21. _____
22. _____

Answers

1. Sciatic nerve, 2. Posterior femoral cutaneous nerve, 3. Saphenous nerve, 4. Tibial nerve, 5. Medial sural cutaneous nerve, 6. Medial plantar nerve, 7. Lateral plantar nerve, 8. Lateral sural cutaneous nerve, 9. Deep fibular (peroneal) nerve, 10. Superficial fibular (peroneal) nerve, 11. Common fibular (peroneal) nerve, 12. Branches from femoral nerve, 13. Femoral nerve, 14. Lateral femoral cutaneous nerve, 15. Lateral femoral cutaneous nerve, 16. Femoral nerve, 17. Obturator nerve, 18. Sciatic nerve, 19. Common fibular (peroneal) nerve, 20. Superficial fibular (peroneal) nerve, 21. Deep fibular (peroneal) nerve, 22. Saphenous nerve

Reference

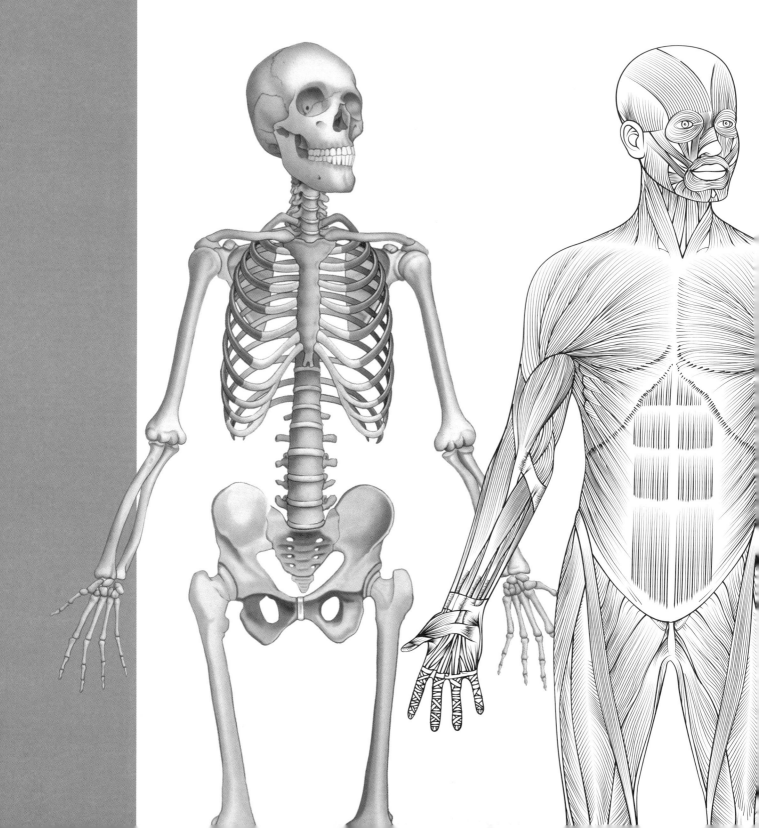

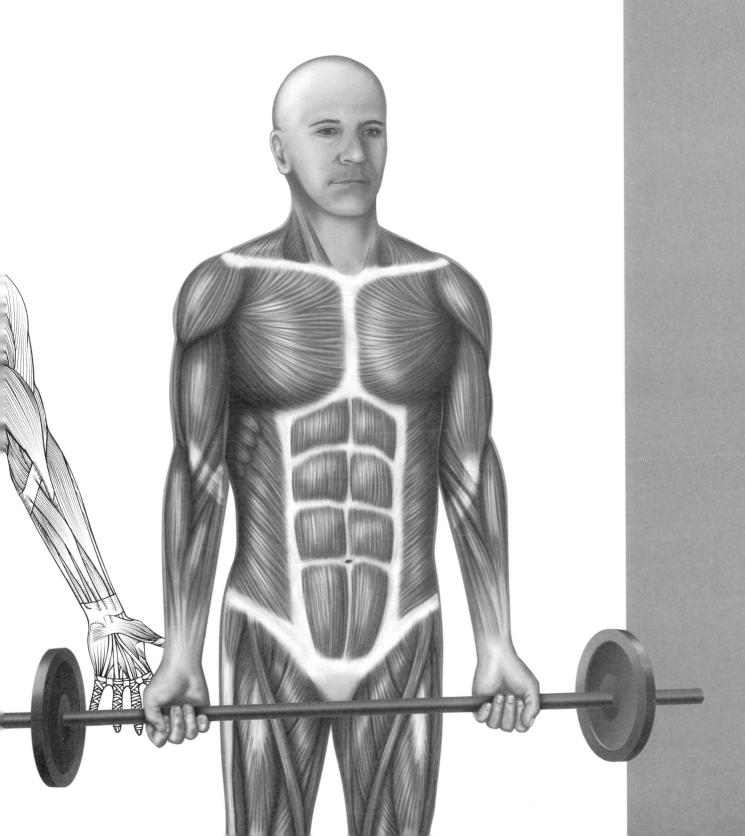

Glossary

Abduction Movement of a limb away from the midline of the body.

Accessory muscles *see* secondary muscles or movers.

Active muscles Muscles responsible for the main movement involved in an exercise. Also known as primary movers.

Adduction Movement of a limb towards the midline of the body.

Alternating grip Grasping the barbell with the palm of one hand facing away from the body, and the other palm facing towards the body.

Bilateral exercise An exercise using both limbs together.

Buttocks (muscles) This group includes the gluteus maximus on the surface, and the gluteus medius and minimus muscles beneath.

Cervical vertebrae The seven vertebrae of the neck. Together they form a curve that is concave to the back of the body.

Concentric contraction A muscle action where the muscle shortens as a result of contraction.

Concentric phase That part of an exercise movement where the muscle shortens as it contracts.

Core The trunk. Often used in reference to core stabilising muscles, e.g. transverse abdominis, multifidus, obliques.

Eccentric contraction A muscle action where the muscle lengthens under tension, such as when lowering a weight against gravity in a controlled movement.

Eccentric control The smooth execution of an eccentric movement.

Eccentric overload The action whereby extra weight is resisted as the muscle lengthens while contracting, such as when lowering a barbell back to the starting position during the bicep curl. The eccentric action can handle more weight than the concentric (muscle shortening) action, so additional weight usually must be added to the barbell if the exerciser wishes to overload the eccentric phase.

Eccentric phase That part of an exercise movement where the muscle lengthens while under activation.

Extension The act of straightening a limb at a joint.

Extensor muscles A group of muscles that perform an extension movement.

Flexion The act of bending a limb at a joint.

Flexor muscles A group of muscles that perform a flexion movement.

Foot strike The impact of the foot on the ground during an exercise movement.

Hamstrings Muscles of the back of the thigh. The hamstring muscle group includes the semitendinosus, semimembranosus, and biceps femoris muscles.

Hyperextension Extension of a joint beyond the normal range of movement.

Internal rotation Rotating a limb towards the midline. Also called medial rotation.

Isometric contraction A muscle contraction where there is no change in the length of the muscle.

Isometrically active Activation of a muscle while staying the same length.

Kinetic chain The connection of all the parts of the body to one another, directly or indirectly. Moving one part of the body can affect the position and momentum of another part of the body.

Kyphosis A curvature of the vertebral column that is concave to the front of the body. It is normal in the thoracic spine.

Lordosis A curvature of the vertebral column that is concave to the back of the body. It is normal in the lumbar and cervical spine.

Lumbar vertebrae The five vertebrae between the bottom of the rib cage and the pelvis. They comprise the lower back.

Medial rotation *see* internal rotation.

Neutral spine The position of the vertebral column or spine, where there is the least stress on joints, ligaments, and discs. In the lumbar spine, this is a position of slight lordosis.

Posterior core Core stabilisers of the back, e.g. multifidus.

Posture During weight training, the correct position for the back is to maintain the normal lordosis curvature of the lumbar spine. Avoid flattening out or arching the back.

Primary mover The main muscle or muscles that produce a movement.

Progressive overload The progressive addition of resistive force to provide optimal training of a muscle.

Proprioceptive neuromuscular facilitation The role of sensory feedback from the stretch receptors in a muscle or its tendon to maintain muscle tone during contraction.

Quadriceps The quadriceps femoris muscle group on the front of the thigh includes the vastus lateralis, vastus intermedius, vastus medialis, and rectus femoris muscles.

Retraction Backwards movement of the scapula towards the vertebral column.

Rotator cuff muscles A group of muscles (supraspinatus, subscapularis, infraspinatus, and teres minor) that arise from the scapula and insert onto the humerus to provide dynamic stability for the shoulder joint.

Scapula The bone commonly known as the shoulder blade.

Secondary muscles or movers Muscles that assist with a movement, but are not primary movers.

Spotter Someone who assists the exerciser to get into the correct starting position, ensures the barbell or dumbbell travels in the right direction, and assists if the exerciser struggles to control the weight during the exercise.

Stabilisers Muscles that are not involved in a movement, but help maintain body position. Important in the kinetic chain concept.

Super set Two exercises performed back to back without rest in between, to create a more intense workout.

Thoracic vertebrae The 12 vertebrae of the chest, comprising the upper and middle back – each is attached to a rib.

Unilateral exercise A single limb exercise.

Index

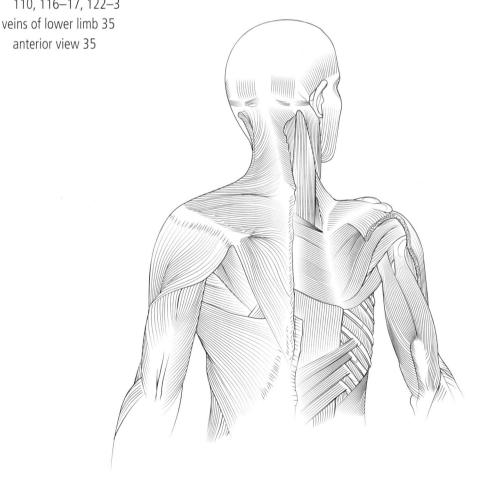

MARR'S GUITARS

JOHNNY MARR

MARR'S GUITARS

CONTENTS

FOREWORD
by Hans Zimmer

The electric guitar is the closest thing to true magic created by humanity that I can think of: not much more than a plank of wood, six strings slung across the neck and body, a few magnets to pick up the vibrations in the strings. And, like nothing else in our world, the elegant simplicity of this design exposes and amplifies everything about the player's individuality. The guitar is an amplifier for the emotional and psychological fingerprints of the soul. If you want to know something about someone, listen to how they play....

How many Fenders or Gibsons are there in this world? And yet even if you want to sound like Johnny Marr, Jimi Hendrix, Nile Rodgers, Jonny Greenwood, Albert Collins, Hank Marvin, Stevie Ray Vaughan, Jeff Beck, David Gilmour, Prince or Tom Verlaine ... you won't be able to. The same guitar lets each one of these artists sound extraordinarily different from one another, allowing each an unmistakable stamp of personality and individuality. Each artist gets their guitar to tell their profoundly personal story.

When I was writing Christopher Nolan's *Inception*, I kept hearing a perfect sound for it in my head. Try as I might I just couldn't identify it. Three days of stumbling around. It was a guitar, but it wasn't one of those guitars you associated with metal guitars and a cheesy orchestra. I wasn't going to give up ... and then, one day I had it. I knew! And I phoned Chris. I remember saying to him 'I have it! I have an idea!' I went on: 'Chris, what do we think of electric guitars and orchestras? Hideous, right? Usually tasteless and pretentious. But ... what if, instead of saying "Guitar", I said "Johnny Marr and orchestra", you wouldn't say "no", would you?'

And Chris understood that it was this player's particular aesthetic amplified by his guitar that I'd been hearing and trying to pinpoint in my head. It was agreed and Mr Marr was on the next flight to sunny Hollywood....

There might be moments when your hero's style and voice shines through early on in your guitar playing – we all stand on the shoulders of giants and we usually try to learn by copying them. But the fun starts when you surrender to your own irrepressible style as it pushes through; when you find the courage to let the guitar speak – shout! – with your own voice. You sound like you. The guitar lets you be you. It's always about so much more than learning scales and chords and just technique. It's a direct connection between you and who you are; a direct expression of all you've lived through, all you've become and all you are at that moment in time. The guitar lets you tell the stories that shaped and formed you.

Guitars tell the truth. The guitar, in the simplicity of its great design, lets us reveal our true selves more profoundly than words ever could.

And think of the aesthetics. The basic shape we are drawn to hasn't really changed that much: the Strat and the Les Paul with their seductive feminine curves; the Fender Jaguar with its almost Dalí-esque variation. Guitars on the whole are sensuous. They 'feel' sensuous. I know there are always exceptions and experimentations but, on the whole, it's the elegance of the organic shape that makes us love the guitar so much: it makes the guitar seem human; it inspires us. And when Woody Guthrie wrote the words 'This Machine Kills Fascists' on his guitar, he acknowledged the power of the thing. Because the moment you pick up a guitar, it gives you a sense of courage and the possibility of talking to the world from a profound place within you – without saying a single word....

Hans Zimmer

INTRODUCTION
by Johnny Marr

I can't remember a time when I didn't have a guitar. I got my first one when I was five years old. I've no idea why I pestered my parents to get me that first one. I just had to have it, and guitars have been with me ever since.

Playing the guitar has defined me, to other people and to myself. It got me through school and my teenage years; through boredom, experimentation and big dreams; through a career and beyond. The guitar has taken me around the world, through fame and fortune. It has given me pure elation onstage. And when I've written a song and hit on a riff, it has also given me comfort and escape when I needed it.

When I started this book, I thought it would be about all the guitars I own. In the making of it, I discovered it was also a story of my life. Recounting why I got a particular model put me right back into the past, and into certain moments. Each day, when I got home after photographing the guitars, I would tell my wife Angie about how amazing it was to reconnect with those emotions, and I would recount how I felt when those particular instruments first became mine.

One question I'm always asked is: 'How many guitars do you have?' The answer is something like 132. That is because 132 was the number the last time I counted. But I don't really know the exact number. Then the person follows up with: 'Wow! That many?' and then without fail asks: 'Do you play them all?' which is actually a good question. The answer is that I do play them, and I've played them all in some capacity or other over the years. Some guitars I've used more than others, and there are a few that I play all the time. But I'm not someone who buys guitars just to sell them on or add them to a collection. I should also say that I'm as amazed as anyone else that I have them all. Wow, indeed!

I wanted to make this book because I love my guitars and I wanted to share their stories. I was also lucky enough to become friends with the photographer Pat Graham, and the combination of his artistry and talent and my beloved instruments was too good an idea to pass up. I also enlisted the talents of my long-time art director Mat Bancroft to add his skills and make it happen. And so, the process of creating this book became a collaboration between me and my brilliant friends.

I hope you enjoy my guitars. I've tried to bring to these pages all the ways that I know and love them. I hope you love them too.

Johnny Marr

GIBSON LES PAUL STANDARD
Black
1980
#82750673

This is the first pro guitar that I owned. I always wanted a Les Paul since I was a kid, because Marc Bolan played one and he was my first hero. I loved the look of the black one and I took the pick guard off it. I bought the Les Paul from A1 Repairs, which was the guitar shop that everyone knew in Manchester. I made the money to pay for it from buying clothes on the Kings Road in London and selling them to my mates in Manchester. At this point, I was in a band called Freak Party. The amazing thing about this guitar is that a year after this photograph was taken, in 1982, I traded it in for a different one, and I'd not set eyes on it for over forty years when a guitar tech friend of mine found it in a shop in Southport. He suspected it was mine, identified it and I bought it back. ■

GRETSCH SUPER AXE
Cherry
1977
NO SERIAL NUMBER

This is the guitar I traded in my black Les Paul for when I formed The Smiths in 1982. I wanted something a bit more unusual, but still classic. This particular model is not the typical type of Gretsch as it has a solid body with an effects unit built into it. The first set of Smiths songs were all written on this guitar, including 'What Difference Does It Make?', 'These Things Take Time', 'Reel Around The Fountain' and 'Jeane'. I used it to record the first single 'Hand In Glove', including all the overdubs, as it was the only guitar I owned at that time. This Gretsch is synonymous with The Smiths starting out, as I played it at all the band's early gigs. In 2022, the guitar became an exhibit in the British Pop Archive at the John Rylands Library, Manchester. ∎

RICKENBACKER 330
Jetglo
1982
#VI2485

The Rickenbacker was the first guitar that I would become associated with. I decided to get one because, as well as loving the look of it, I thought it might make me write and play in a certain way. Unlike Gibson Les Pauls or Fender Stratocasters, the Rickenbacker isn't known for a traditional rock approach and I thought it would be good for my melodic style of playing, focusing more on chord changes and hooks. It sometimes sped my playing up a bit too, as quite a lot of the songs I wrote on it were fast and hyperactive sounding. The best example of the way it sounds is on the intro to 'What Difference Does It Make?' by The Smiths and on 'For You' by Electronic. I still go back to this guitar and have played it on lots of things right throughout my career. It's also pictured on the cover of the Oasis single 'Supersonic'. ■

GIBSON ES-355
Cherry
1960
#A32530

In some ways, this is my most famous guitar. It was bought for me by the head of Sire Records, Seymour Stein, on 2 January 1984. I was a big fan of Sire, and when I discovered Seymour was interested in signing The Smiths I was thrilled about it. He took the band out to dinner before we played at the ICA in London, and during that evening he told me that in the 1960s he had bought a guitar for Brian Jones of the Rolling Stones. I told Seymour if he bought me a guitar we'd sign to Sire, even though we were going to sign anyway. The band headed out to New York to play our first American show at Danceteria on New Year's Eve 1983. A couple of days later, after going into the Sire offices to sign the record contract, I marched Seymour through the snow to We Buy Guitars on 48th Street. Seymour was true to his word and bought the guitar for me. I raced back to The Iroquois hotel, where the band was staying, took the guitar out of its case, and the first thing I played on it was what would become the next Smiths' single, 'Heaven Knows I'm Miserable Now'. The song just arrived under my fingers, complete. Then the very next thing I played was what would become the B-side, 'Girl Afraid'. Some instruments come with songs in them, and this guitar came with quite a few. It coincided with the band moving to another level of success, and when people saw me using it on *Top of the Pops*, it made an impression. Noel Gallagher and Bernard Butler went on to use red Gibson 355s in their own bands, Oasis and Suede. The guitar also became well known for a photo that was taken after a gig at the University of East Anglia, Norwich, in 1984. ■

FENDER TELECASTER
GIFFIN CUSTOM KORINA
Green Burst
c. 1984
NO SERIAL NUMBER

Angie bought me this Tele for a present in 1984. It was built by the master guitar builder Roger Giffin, and it's another guitar that got a lot of attention when people saw it on TV because of its unique appearance. It's one of the guitars I am asked about most often, and it's become a part of my history. ∎

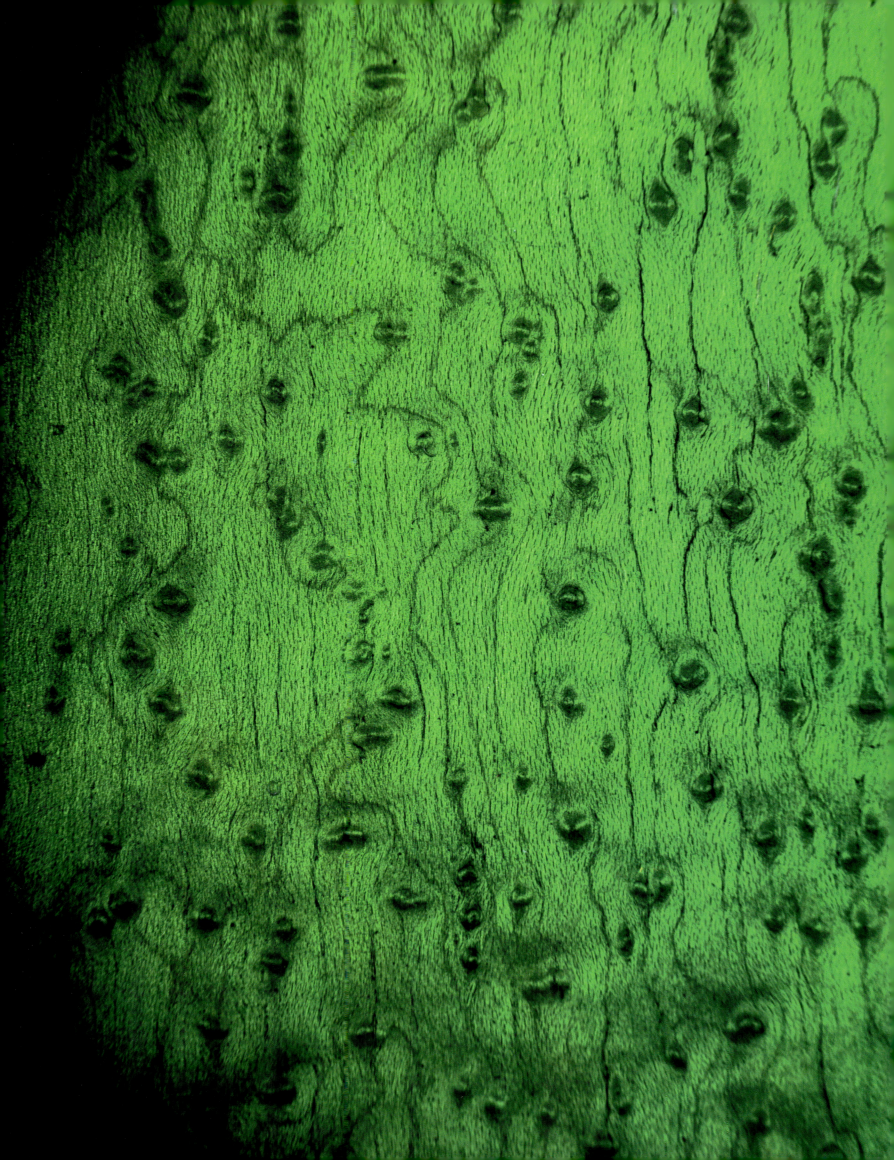

GIBSON J-160E
Sunburst
1963
#148820

I've always loved playing acoustic guitar, but this was the first decent model I owned. I got this Gibson because of my love of the Beatles and the Hollies, and I'd discovered that a lot of the beat groups of the early 1960s used this model. I'd moved into a flat in Earl's Court and every night I'd watch a video called *The Compleat Beatles* with footage of John Lennon playing his J-160E on songs like 'I Should Have Known Better' and 'Things We Said Today'. This guitar inspired me to write 'Please, Please, Please Let Me Get What I Want', 'William, It Was Really Nothing' and 'Well I Wonder'. It brought a bit of a beat group sensibility to my writing. ∎

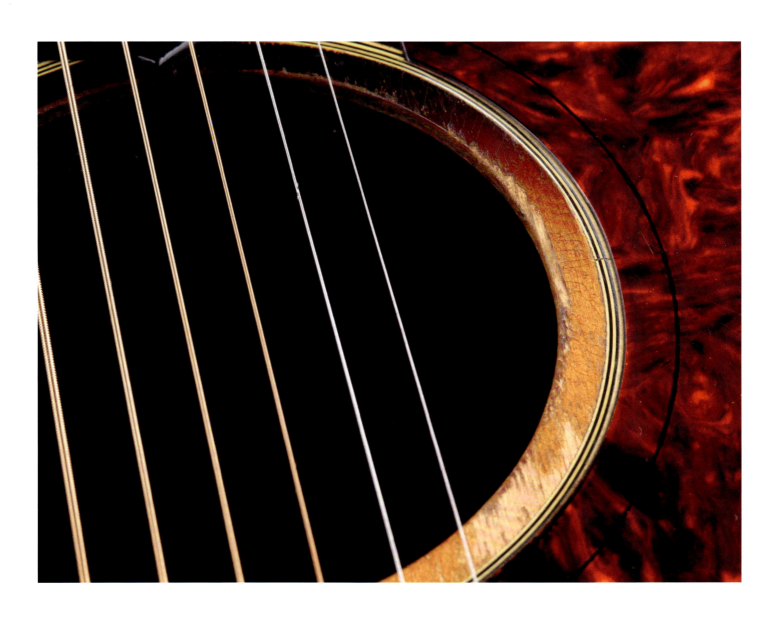

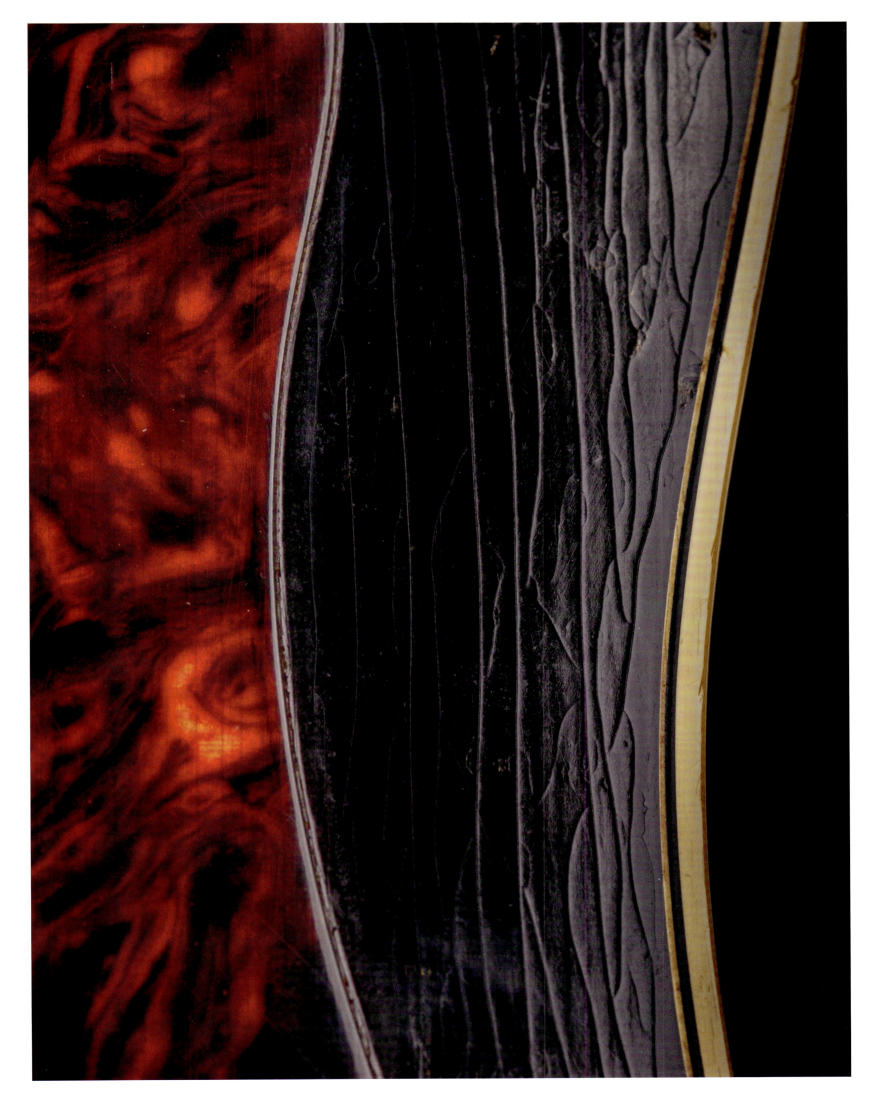

EPIPHONE CORONET
Cherry
1962
#49281

This is one of many guitars that I bought from John Entwistle from The Who. I set it up to do what's called 'Nashville tuning', which is when you replace the bottom four strings with the strings from a 12-string set that are an octave higher in pitch and leave the top two strings at regular pitch. The sound of Nashville tuning is like a 12-string guitar, but brighter and more focused and delicate. It can be heard very clearly on the songs 'William, It Was Really Nothing' and 'The Headmaster Ritual'. It's a technique I mostly use for overdubs as it's effective and doesn't get in the way of the other guitars on the track. ■

OVATION LEGEND 1867
Sunburst
1985
#336054

I needed to have an acoustic guitar to use at shows to replicate the acoustic sound I was using on the records. The best guitar for that in the late 1970s and early 1980s was made by Ovation. I also like the way these guitars make me play because they feel a little bit like an electric guitar while making the sound of an acoustic. ■

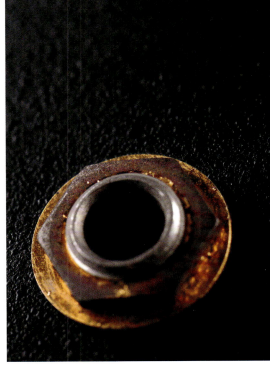

INTERVIEW: PART 1

Johnny Marr with Martin Kelly

MK I remember hearing you with The Smiths for the first time in 1983 and being transfixed by the guitar playing and the sound of the band. It was completely fresh and unlike anything I'd heard before. Stylistically, I couldn't pin it down, but I knew straight away that this was someone special playing the guitar. I think it's safe to say that you redefined the term 'guitar hero' for my generation and several that have followed, but what drew you to the guitar, what was its allure?

JM I think it was something about the shape. As a little boy growing up in the 1960s, I can only equate it with my mates who wanted a bow and arrow or toy cars. I just wanted this little wooden toy guitar, which was probably about this big [gestures around 2 feet/60 cm in length]. I was five when I got my first guitar from a small shop that sold mops, buckets and brooms. My parents say that whenever they tried to walk past the shop I was glued to the window, and my mother had to pull me along. My mum would say to my dad, 'He's besotted with this guitar in this shop.' I remember the shop and I remember the moment the guitar was taken out of the window and handed to me, in a little box. My mates were running around playing cowboys and I carried my little wooden guitar around, the way they carried toy guns and bows and arrows around. It really was like my best friend.

There was a connection with music because our house was so full of it. To understand that, you have to know that my parents, who are Irish and in their seventies, they'll be listening to music right now. My mum may be watching some singer-songwriter she likes on YouTube, and my dad will have taped something off the radio. That's the kind of house I grew up in, absolutely obsessed with music, to the point where it was like a religion.

As well as the parties and get-togethers we had in our house with aunties, uncles and my little cousins, my parents were also going out to the Irish clubs. There was an Irish community in Manchester in the early 1960s, clubs called The Ardri and The Carousel. Showbands would come over from Ireland, and there was this whole network of venues for them to play –

Manchester was really big for those showbands. My parents would see people like Johnny McEvoy, Big Tom and the Mainliners or Joe Dolan. So, I was very used to my parents – particularly my mother – being very excited about seeing a band. Maybe even on a Wednesday night, she would come in a little buzzed with her autograph book, saying, 'John, John, the guitarist ....' So, that was the atmosphere around me. And maybe that's what made me understand what a guitar was – whereas my pals, who were into footballs or toy cars or trains, maybe they didn't understand it so much.

My first experience of seeing live music was a band that used to play at christenings, weddings and parties in a room above a pub. You'd have the event in the church and then everyone would bowl down to the pub, and me and my cousins would be running around. The function room upstairs would be empty at that point and I'd hear the DJ setting up. I remember he'd be playing ska records or The Equals, stuff like that, late 1960s pop music, setting up his decks. So, my ear was immediately like, 'Right, I'm going up there.' I would have been about seven or eight, and I would head upstairs and be checking out the DJ on my own in this completely empty room. And then the band would arrive and start lugging their gear up the stairs onto this little stage and I'd watch them setting up.

I remember the second time I watched them, I was waiting for the guitarist to pull out his red Strat. It occurred to me then, 'Ah ... this is their job.' They would have been in their mid-thirties, lugging their gear in and setting it up. I could actually see how it was all put together, before they started rehearsing 'Love Grows (Where My Rosemary Goes)' or whatever they were going to play. So, I'm a few feet away from the guitarist, taking it all in, and when the band finished their set my dad went over and said, 'He's mad on the guitar, can he have a look?' The guitar player had put his guitar away, but said, 'OK.' He was an Irish guy, and when he opened the case, it really was like *Raiders of the Lost Ark*, *Pulp Fiction* or one of those kinds of incredible moments where this artefact, this holy object, was presented in front of me. I remember it was a red Strat, and the smell of the case and all of that.

There was nothing more magical to me. I liked football, I liked clothes ... you know, I was a little kid, so I liked cartoons on the television. All the stuff that regular little boys like. But my toy guitar, the experience of seeing that first band and the things I watched on mainstream television (you know, 'showbusiness' and what was called 'variety' back then) – that was a different part of life for me. Everything else was 'normal', but that other stuff was magic, and I've never really lost that. It's what I do and what my mates do. And I don't just mean my famous mates, I mean some people in the record industry. I see it all as being a magical realm really.

I've always been drawn to music that features a guitar hook. It goes back to seeing adults absolutely rocking in our front room to Eddie Cochran's 'C'Mon, Everybody'. When I heard that, in say 1968 or 1969, it was already a retro record. My parents were still into rock 'n' roll, especially if they'd had a few drinks. With 'Don't Be Cruel', even though you've got someone like Elvis Presley singing on it, the guitar is the first thing you hear, and the guitar is the star as far as I'm concerned.

Luckily for me and for some of the bands I've been in, I still regard the guitar as the star. Even if I'm in a band with a great singer – and I've been in several now – it's my job to think that the guitar is the star. A lot of my favourite pop records when I was young were built on absolutely stonking guitar riffs. If we're talking the late 1960s, you've got some great riff-driven Motown records, but when I really started to come into my own was the glam rock era. I was probably about ten or eleven, and was into all those glam rock groups like The Sweet and T. Rex. The first 45 I ever owned, I bought purely because of the label, but it just happened to have a fantastic blues riff and it was built on a great guitar part. And that was the stuff that was important in my life.

But before I discovered Marc Bolan, there was a time between the ages of six and nine when I used to stand in front of the radio and just stare into it for an hour or two at a time. My mother would stand me up on a chair and I would stare into the radio and listen to what was coming out, regardless of what it was. It was just the pop music of the day, and luckily for me there was a lot of good pop music around.

This was the late 1960s, so there was a lot of Clodagh Rodgers, Engelbert Humperdinck, Des O'Connor and all this real mainstream stuff, what the Germans would call 'schlager' music. But I was also hearing a lot of Motown, the Hollies – who were very important – and, of course, the Beatles. It being the late 1960s,

you couldn't escape the Beatles. So, I was obsessed with what I was hearing on the radio.

The only way I could actually see guitars at that time was on the television. There would be a mainstream comedian or *Saturday Night at the London Palladium*, what you would call 'variety performances', and if it was the Cilla Black show or the Lulu show, I'd wait for the musical segment and I'd just hope it was a band. I didn't really care who it was, as long as it had a guitar or a bass. I'd see these mainstream acts like Blue Mink and one of them would be playing what I now know is a Fender Precision Bass, and I'd recognize that the same kind of bass guitar kept popping up. Quite often, the lead singer would be playing an acoustic and then you'd have an electric guitarist, too. One of my favourites at that time (because you really didn't see the Beatles on those kind of shows) was the Hollies. They always had good guitar parts and I liked the guitarist Tony Hicks. He looked great, and what he was playing was great – and it was wrapped up in pop music. So, those singles like 'Bus Stop', 'I Can't Let Go' and 'On A Carousel', they were just magical things to hear on the radio or see on television.

Around this time, my mum would take me with her when she went to the department stores in Manchester. The first thing she would do was to stand me where they sold the guitars, and then she would go off and do her shopping. Because that was where I wanted to be and she knew that I would be OK there. Different times. There was a point over the years when I had to really check this, because it was coming up in interviews. So I asked my mother, 'Are you cool with this? Did this really happen?' And she said, 'Oh yeah, and it wasn't just the guitars, it was the amps too.' So, looking back on that, I really was completely obsessed with what the guitar was doing and its role in the music I was hearing.

Luckily for me we lived within a ten-minute walk of the inner city, or 'downtown' as it's called in the USA. Me and my sister were pretty savvy, very streetwise for our age. One of the things I remember – I don't know whether it's an Irish thing – but we were given a lot of leeway to roam free, and we were very independent. So town, the city centre, was of great interest to me.

We would go into town all the time, not just on the weekends. That's when I started to notice the guitar shops, which were like treasure troves for me – whether they were open or closed. I used to go in quite a lot on a Sunday because there wasn't much else to do. I'd walk into town down Oxford Road.

MK In early 1972, as part of an inner-city clearance scheme, you moved with your family from Ardwick to a brand-new housing estate in Wythenshawe. It's only 7 miles (11 km) south of Manchester's city centre, but you described the change as 'like we'd moved to Beverly Hills'. You suddenly had a new school and new friends. It sounds like that move really helped you grow into life as a musician. There were a number of guitar shops in Manchester in the early 1970s, and one of them in particular was the 'go-to' shop for everyone, including Manchester's professional musicians. It was called A1 Repairs. There was one called Mamelok; there was also Forsyth, which is still there, and Barratts was another.

JM My first proper guitar, one I could tune and play chords on, was an acoustic that I got when I was eight. I got it from Barratts. Although it was a fairly cheap guitar, it was still a luxury item. It was the first time my passion was recognized by my parents, and they really liked getting me that guitar. It wasn't like a skateboard or something that they thought I would be done with in a year. They were like, 'There you go, that's what you really want and that's what you do.' When we were going home with it, I sensed that my dad – who was a man of very few words – knew it was a big moment for me.

Marc Bolan crashed into my life when I was about ten and I bought 'Jeepster' by T. Rex. He was my first proper hero. He was small and beautiful, and he played a great Les Paul and also a white Strat. The first songs I learnt to play on my acoustic were 'Jeepster' and 'Stay With Me' by the Faces. But the first song that I really thought I could play properly was the B-side of 'Jeepster', which was called 'Life's A Gas'. Until then, I was learning chord shapes and putting my own chord shapes together. But 'Jeepster' and 'Life's A Gas' made me want to learn how to really play, and luckily they used chords I already knew. I also got a two-page sheet from someone that had the chords to 'Let It Be', which were G, D and E minor. That's the kind of thing I was learning. But pretty much as soon as I could play D minor and F and that kind of stuff, I started to make up my own little tunes. It became the main thing for me to do. It still is.

I got my first electric guitar after I managed to blag a job in a little guitar shop in Altrincham, a small town in south Manchester. I would go in there every Saturday morning and chat to the boss. I think he was amused by me, because I would have been maybe eleven or twelve,

and I just hung around asking him questions about the guitars. One day, I asked him outright, 'Can I work here?' He soon started to twig, after the fifth or sixth week of me turning up, that I was going to be in there for hours anyway, so he might as well have me doing something. He couldn't legally employ me, so he offered to give me a discount on a guitar in exchange for me running around for him and moving boxes. The cheapest guitar in the place was a red second-hand Vox Ace, a vaguely Strat-shaped three-quarter-size thing, but it was a proper electric guitar and it was just about affordable. So, we had this arrangement where if I made a certain amount of money – £15, £20 or whatever it was – I could get a discount on the Vox Ace. Like a lot of kids, I got a paper round to fund the difference, so the guitar cost me about £35 in the end. I actually got going on that guitar and put it to good use – I was off. I learnt all the riffs from the records at the time – things like Bowie's 'Rebel Rebel'. It only occurred to me after being asked about these riffs years later that the start of the very first Smiths single sounds a bit like 'Rebel Rebel'. So, it all goes in; it all becomes part of your technique, your schtick.

My big problem at the time was not having enough money for an amp. There was no way I could even imagine getting one. I had to find mates who might know someone who had an amp and would let me plug into it for half an hour in their bedroom. I did that plenty of times.

I was at a boys' grammar school and was mixing with older lads, or boys who had older brothers. On a few occasions, I would get one or two buses, or a bus and a train, and turn up at a boy from school's house because his older brother had a guitar and an amp. I would just turn up, even if he didn't really know me. Back then, guitars weren't nearly as common as they are now. They were luxury items and were pretty rare. We're now in an age – and have been for decades – where everyone knows someone who has a guitar. One time, they were real rarities. I knew of every guitar that was within a 10-mile (16-km) radius of where I lived – the name of who owned it and what they had. There was a guy who lived on my estate who had a sunburst Telecaster. I knocked on his door a couple of times when he was watching television and just said, 'Can I see your Telecaster?' I remember him because he had that guitar. There was also another guy I knew whose older brother had a guitar. I had to get two buses to go to his house. The family would be having their tea and it was a major inconvenience when this Wythenshawe boy, who they didn't really know, turned up to say, 'Can I see your brother's guitar?' It's pretty funny now that I think about it.

That was what was going on until I found out about another young guy in the neighbourhood called Kevin Williams, who had a Vox AC30 amp. Back then, this was the equivalent of owning a Cadillac. We started knocking around together and the AC30 became my first band's amp. We would all plug into it, including a microphone for Chris the singer. I roped in Kevin because he had the amp and he could play pretty well. He later found fame as an actor in *Coronation Street*.

●

JM I loved all guitars but started getting together my own taste, my own individuality, aged around fourteen or fifteen, and that coincided with punk. I always wanted a Les Paul, which went back to liking Marc Bolan. He really put it into my mind that the Les Paul was the most desirable guitar. But then a lot of the punk guitar players that I started to see were also playing Les Pauls. The first gig I ever went to was a local band called Slaughter and the Dogs in south Manchester. I was twelve when I saw them in 1976. The guitar player in that band, Mick Rossi, was playing a white Les Paul. So, I kept seeing Les Pauls, and they were in my mind. But I had to borrow guitars and amps off mates for quite a while until I got my first Les Paul. I took a blowtorch to the Vox Ace to try and make it look like Rory Gallagher's beaten-up Strat and completely ruined it – obviously. So, then I had to sand it down to a natural wood finish. Eventually, I got together enough money to buy a sunburst Satellite Les Paul copy. That did me for a couple of years, and I started to get good on that. I'd had a couple of bands by that time and then I joined my first professional group, Sister Ray, who were playing shows and had made a record. I was still playing my Satellite Les Paul then, which was as close as I could get to James Williamson's 1973 Les Paul Custom.

After hearing *Raw Power* by Iggy and the Stooges, and with my love of Marc Bolan, no other guitar was going to compete with a Les Paul. The first guitar in this book is my black Les Paul Standard and I got it when I was seventeen. I was with my girlfriend Angie, who's now my wife, and we'd got into the habit of going down to the King's Road in London and buying clothes wholesale, then bringing them back to Manchester and selling them to our mates. Eventually, the two of us

saved enough money to buy a second-hand Les Paul Standard and a Fender Twin Reverb amp. That was the first guitar I ever bought from A1 Repairs in Manchester. I would end up buying quite a few important ones from that shop over the years.

My best pal for a long time was Andy Rourke, who later became the bass player in The Smiths, and me and Andy would jam for hours. Me getting the black Les Paul coincided with Andy switching to bass, and I was coming up with a lot of ideas for songs and riffs. Andy turned me onto a lot of stuff – The Velvets and Eno. We both played in a band with older guys for a while, but then there was this whole explosion of more interesting music that made us leave that band. And then we started getting into our own kind of thing.

So, we're talking 1979, 1980 now and I formed my own band, with Andy on bass, and we were doing a lot of power pop stuff: 'Do Anything You Wanna Do' by Eddie and the Hot Rods, 'Hanging On The Telephone' by Blondie, 'Suffragette City' by Bowie, all that kind of thing. That was my apprenticeship as a writer and a guitar player, playing other people's songs. It was a good learning experience. I went through a whole thing with acoustic guitars as well of course; that was very important. I felt I had to be as good on acoustic as on electric. So, I started building a repertoire and developing my own style, even though I wasn't totally aware that I was doing it. I just knew I wanted to have my own thing that wasn't copying anyone else.

I think I did have a feeling that I was onto something. I'm still a bit like that now. It's what keeps me going. There's something I want to invent that hasn't been done. That's what I want to do. I remember having that feeling because I told all my mates, 'I'm disappearing for a bit'. And I had a year zero, where I completely isolated myself. There was just me and Angie and the black Les Paul, and I went back to my parents' house and back to the drawing board. I treated my bedroom, my record player and my Fender Twin Reverb like a laboratory. I used a TEAC tape machine, which you could overdub on by just plugging into the front of it. I took all these concepts that I had and just did it all on the guitar, with no drum machine or effects. If I wanted to put a beat down, I would do it by scratching rhythm on the strings. That would give me my drum sound. Then I got a Boss Flanger pedal. I started experimenting, trying to find what I was feeling like, and those experiments turned into the early Smiths songs.

GIBSON LES PAUL STANDARD
Cherry
1984
#82264538

I bought this guitar from A1 Repairs in Manchester for the second Smiths album *Meat Is Murder*. It may surprise some people to learn that this is the guitar I've used on more records than any other. When people think of me having a ringing sound, they usually assume it's a Rickenbacker, Fender Telecaster or Fender Jaguar, but it's often been this 1980s Les Paul. I put the Bigsby trem arm on myself because I liked the look of it on Neil Young's old black Les Paul. It's the guitar I played on the last song The Smiths played together in concert. I used this on 'Slow Emotion Replay' and 'Dogs Of Lust' by The The. It's also the guitar Bernard Sumner used on the New Order song 'Regret'. ■

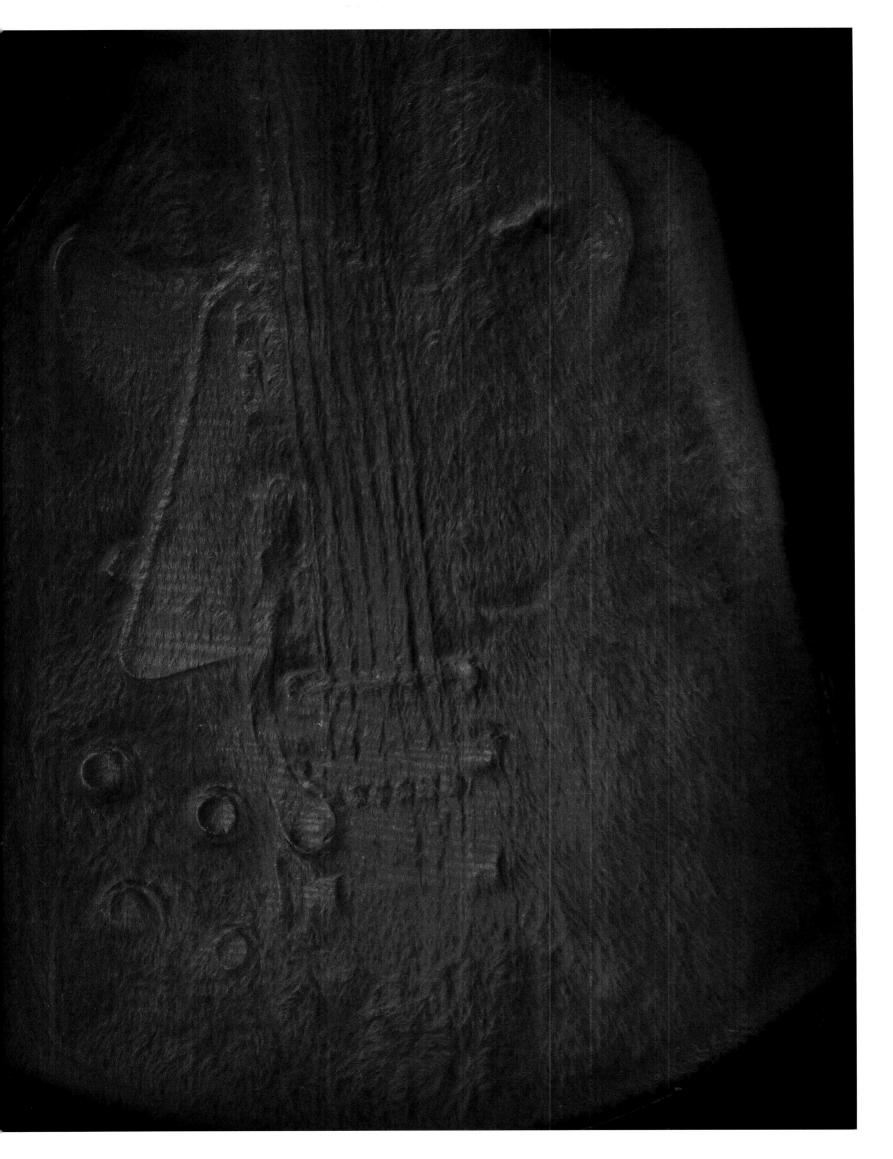

MARTIN D-28
Natural
1971
#286524

I bought this Martin in 1984 when I could finally afford a proper classic acoustic. I could only dream of owning a guitar like this when I was a kid. To me, this guitar is synonymous with people like Joni Mitchell, Neil Young and Bert Jansch. I've always felt that if you don't learn to play acoustic, then you're only living in half of the guitar world. It was a conscious decision to feature acoustic guitars on The Smiths records – and have them loud whenever possible. This Martin became my main acoustic for years, and it's the guitar I used to write 'There Is A Light That Never Goes Out' and 'Cemetry Gates'. ■

EPIPHONE CASINO
Sunburst
1963
#140534

Another classic beat group guitar that I got from Denmark Street for its Beatles, Kinks and Stones connections. It looks great and has got a unique sixties sound. I wasn't able to use it very much live because it would feedback quite easily at high volume, but I did use it at Glastonbury in 1984 and wrote 'Nowhere Fast' on it. The main thing about this guitar is that it's the one I used for the tremolo riff on 'How Soon Is Now?' ∎

TWIN REVERB®

MASTER VOLUME

FENDER MUSICAL

VIBRATO

1

2

FENDER STRATOCASTERS
Olympic White
1962 / 1962
#70012 / #86694

Always, I hoped that my latest guitar would make me do something new, as the Strat on the left did when I wrote 'The Boy With The Thorn In His Side' on it. It was the first one I got and was formerly owned by Gary Shaughnessy from the Manchester pop band Sweet Sensation, known for their Number One song 'Sad Sweet Dreamer'. I liked it so much that when I came across another one soon after, I got it as a back-up, and eventually that one became my main guitar with The Pretenders. ∎

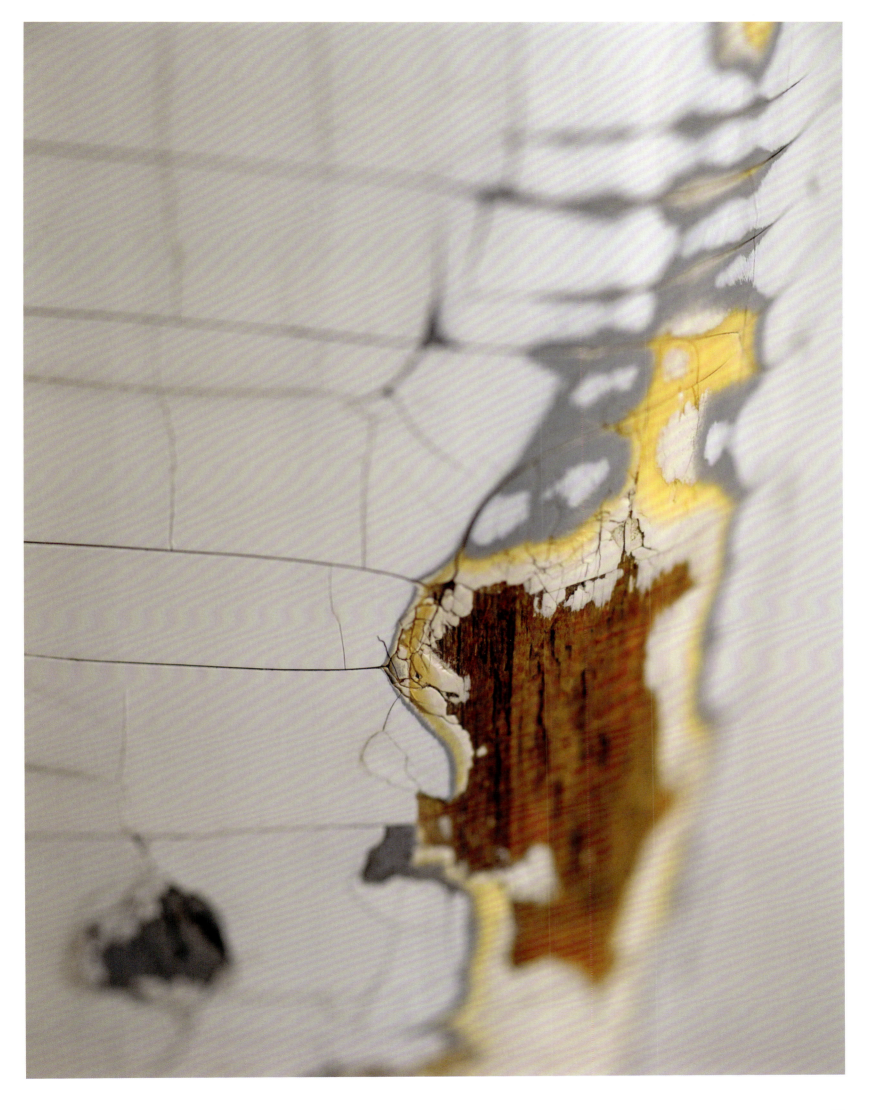

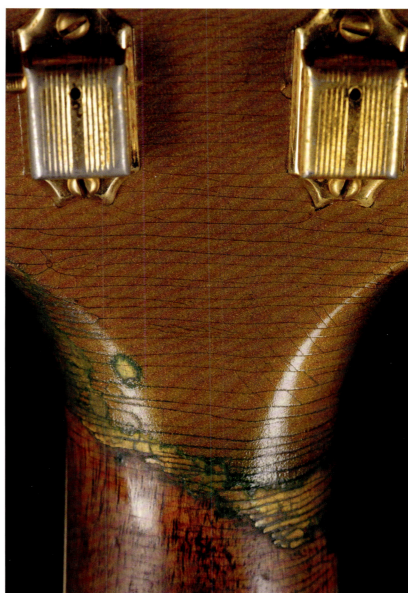

GIBSON ES-295
Gold
1952
#A12411

In the 1980s, guitars from the 1950s and 1960s weren't referred to as 'vintage'; they were just called 'old guitars'. There were a few of us younger guitar players around London in the 1980s who were into playing these types of old guitars, and we would occasionally bump into each other in the couple of shops that sold them – people like Roddy Frame from Aztec Camera, Edwyn Collins of Orange Juice and Robin Guthrie from Cocteau Twins. I lived around the corner from a place called Earl's Court Guitars, which would eventually become New Kings Road Vintage Guitar Emporium. I bought this Gibson and quite a few other things from my friend Rick, who was the manager. I had a rule that if I bought a guitar, I would have to write a song on it to justify the expense. When I saw this one, I loved the look of it. To me and most other guitar freaks, the gold ES-295 will always be associated with Scotty Moore who played one with Elvis Presley. ∎

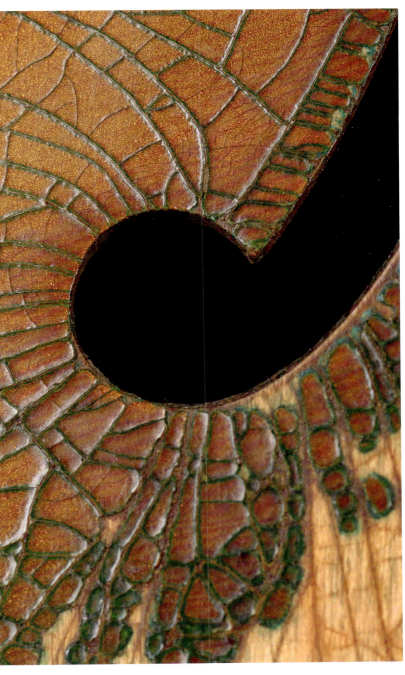

GRETSCH 6120 SINGLE-CUTAWAY
Western Orange
1960
#38802

I liked that people like Eddie Cochran and The Cramps would use big hollow body guitars like this one. They remind me of those great sounding rock 'n' roll records from the 1950s. I got this one from John Entwistle of The Who and used it on 'Ask' by The Smiths and in the video for 'Panic'. ∎

GIBSON LES PAUL STANDARD
Sunburst
1953 [converted to 1958–60 spec]
NO SERIAL NUMBER

Another guitar from The Who. I used this onstage and in the studio from 1986 onwards. In 1993, my brother Ian told me about a friend of his who was in a band and who he thought that I might like. Shortly afterwards, we were driving through Manchester and saw this figure in a duffle coat, walking in the rain. I pulled over to give him a ride, and that was the first time I met Noel Gallagher. We went for a cup of tea and immediately got along. He struck me as someone who was very dedicated and on a mission. We met up again a few days later to go to a guitar shop, and around the same time I went to see his band play in front of about fifteen people. Then a couple of weeks later, I went again, and this time there were about twenty people.

Soon after we were talking on the phone, and he asked what I thought of the shows. I told him they were great, but I noticed that he seemed to be taking a long time tuning up his guitar between songs and that it would be a good idea to have a second guitar. His response was: 'That's alright for you. I'm on the dole so I've only got the one guitar.' This made sense and gave me something to think about. I went into my studio to look for a guitar that he could use. I didn't think it was right that I should offer him one that wasn't great, so I sent this Les Paul down to the studio he was in. I think he was impressed that it was one that I'd used on Smiths records and that I'd got from The Who. He immediately started using it, writing the song 'Slide Away'. As Oasis started to take off, he used it for the band's national TV debut on *The Word* and then in the video for 'Live Forever'. As soon as I saw him playing it, I realized the guitar should belong to him. Over the years, the story of me giving Noel the guitar has been told in many different ways. The one that I like most is that he and I met on a grassy knoll under a moonlit sky at midnight and I passed over the Les Paul to him, saying: 'Here ... Noel of Burnage, taketh thy Les Paul, that thou may go forth into the world and lay down some heavy licks', and we then drank the blood of a groupie. It's a good story, and I've always been happy that such an important guitar went to such a good home. To be continued...

'Oasis arrived at Monnow Valley Studio to record what would be an aborted attempt at "Definitely Maybe". Having only two guitars and a bass between us, Johnny asked would we like to borrow some of his for the recording. Of course, we said yes, and a couple of days later a van arrived. There was this black and white Ricky, which when I saw it, I remember thinking that can't be THE black and white Ricky? Surely he's not THAT insane to send off into the night one of The Smiths' most famous guitars ... the one that created the ACTUAL sound of The Smiths? Turns out he IS that insane. I opened the other case, which contained the Les Paul ... it was a moment not unlike Vincent Vega opening Marsellus Wallace's briefcase in Pulp Fiction ... having never actually seen a "real" Les Paul before, I kinda stared at it for ages. I remember making a cup of tea, taking it to my room upstairs and taking it out of the case. I literally wrote "Slide Away" in about twenty minutes. It was a pretty freakishly magic moment. One I'll never forget.'

NOEL GALLAGHER

GIBSON LES PAUL CUSTOM
Black
1978
#73048576

... One morning in August 1994, someone from my manager's office (which also managed Oasis by then) called me to say that there had been an incident at the Oasis show in Newcastle the night before, and the sunburst Les Paul had been damaged. They asked if I had another guitar Noel could borrow temporarily. I ran into my studio to see what I could come up with. It didn't seem right to replace a very cool guitar with something inferior, so I grabbed this black Les Paul – which was the other main Les Paul that I used onstage and in the studio from 1985–87, most notably for *The Queen Is Dead* album – and took it to Birmingham for their next gig. I put a note inside the case that said, 'This is a heavier one. If you get a good swing on it, you can take some fucker's head off.' A few days later, I called him to say he could keep it. I would play it again eighteen years later on the Oasis album *Heathen Chemistry*. ∎

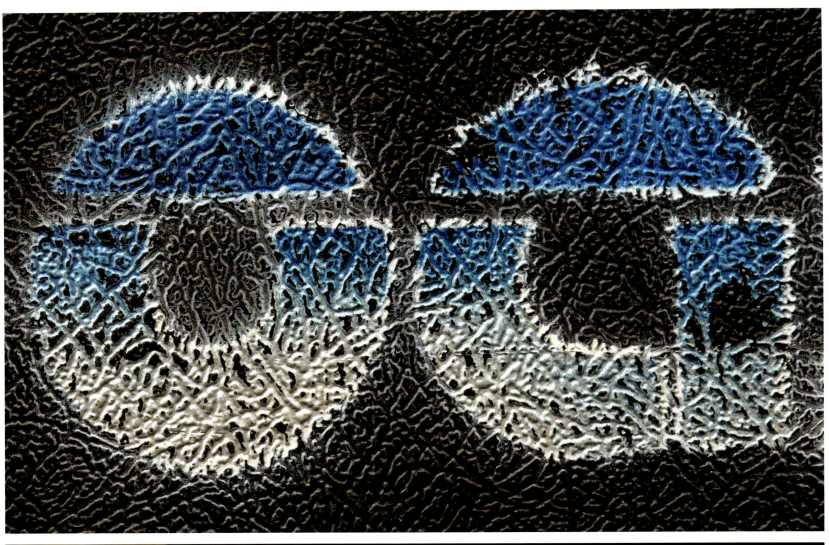

'We were about thirty minutes into a gig in Newcastle, when some lad got up onto the stage and started a minor fracas that turned into a brawl. During the ensuing chaos, he might have got himself clattered on the head with what has now become a very famous guitar! He should be very proud. I'm not exactly sure how, but the sunburst Les Paul got damaged at that gig in Newcastle ... clearly, it wasn't designed to be wielded like an axe. Johnny being Johnny turns up at the next gig (in Birmingham) with the black Les Paul. Now something happened to that guitar a few years after he gave it to me that caused me to swap the pickups out for a set of P90s from a 1968 Firebird (the one I used on the Oasis song "Morning Glory"). Those are the pickups that are on it now. About twenty-four years later, he casually tells me that the Les Paul was THE The Queen Is Dead guitar! I'm not sure I ever actually wrote anything on it, but I did use it on the Oasis album Standing On The Shoulder Of Giants. Johnny used it on Oasis sessions for Heathen Chemistry, definitely on "(Probably) All In The Mind". I do remember him asking though, "Are these the same pickups?"'

NOEL GALLAGHER

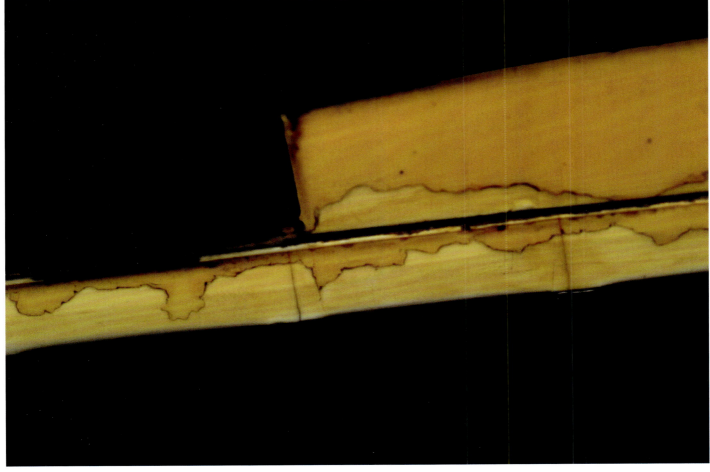

FENDER STRATOCASTER
Sunburst / Maple neck
1963 [body] / 1957 [neck]
#66072

Along with the Les Pauls and Gretsch 6120, this guitar was a big part
of The Smiths' 1986 US and UK tours and *The Queen Is Dead* album. ∎

SMITHS PEDALBOARD
BOSS BCB-60
CS-3 / CE-2 / DS-1 / OD-2 / GE-7 / DD-2

My Smiths pedalboard. I didn't use a lot of effects in the 1980s, but I needed to reproduce some of the sounds on the records I was making. Eventually, I would employ more expansive sounds with technology that I would refer to as 'producing with your feet'. ∎

GIBSON ES-335/12
Sunburst
1968
#905064

I use 12-string electric guitars a lot, and these models are great because the pickups are more rock orientated, adding a darker sound than you might get on other 12-string models. I've been using these guitars for so long that I'm always surprised at how uncommon they are. This one was a big part of the *Strangeways, Here We Come* album, particularly on songs like 'Paint A Vulgar Picture' and 'Stop Me If You Think You've Heard This One Before', and I also wrote the single 'Sheila Take A Bow' on it. After that I used it with Talking Heads when they invited me to record with them in Paris in 1987 for their album *Naked*. This guitar now belongs to my friend Bernard Butler, who I gave it to in 1995. ∎

'I look after Johnny Marr's Gibson ES-335/12. It sits quietly next to some fine semi-acoustic brothers and sisters, waiting for its moment. It has done since 1995 when, at the end of a blurry weekend in Manchester, Johnny produced a case from the loft and made sure it went home with me. It's Johnny's guitar, and while I look after it, I am in debt for the beautiful sounds that ooze out of it, as it would for him.

Nobody picks a 12-string first because nobody wants to tune them. You need to really earn your part for this to sing. Not overly bright on the bridge, probably perfect on the middle pickup, it sweetly arpeggiates and rings with joy when strummed. It is reserved for moments of magic. Moments of Johnny magic. My favourite game? To hand it unbeknownst to some young whippersnapper, dial up "Shoplifters..." from The Tube and watch jaws drop. "Do you know what this one is? – It's that one."'

BERNARD BUTLER

FENDER BASS VI
Sunburst
1962
#76058

An unusual and useful instrument. This is sometimes known as a 6-string bass, but I've always called it a baritone guitar. I have the strings tuned to a pitch between the bass and the guitar, not as low as a bass or as high as a guitar. The strings were custom made for me by Rotosound, replicating the ones they originally manufactured for John Entwistle in the 1960s. ■

FENDER STRATOCASTER
Sunburst
1965
L68296

'The Strat ... now that guitar has been played on some MONSTER tunes. So, we were getting ready to go and record (What's The Story) Morning Glory?, *and I was attempting to spread my wings guitar-sound-wise. Johnny said he had a Strat he was selling. (I think at this point he'd got sick of "lending" me guitars.) I used it on "She's Electric", "Don't Look Back In Anger" AND "Wonderwall". I put it back in its case and never even saw it again until 2010, when for some reason I had it at home and wrote 'AKA... What A Life!' on it. It's on the recording too.'*

NOEL GALLAGHER

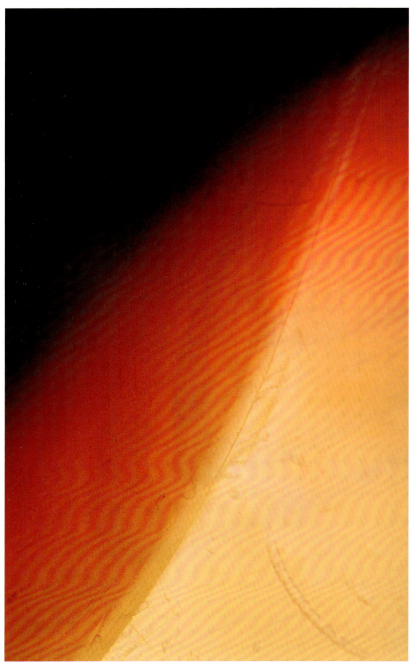

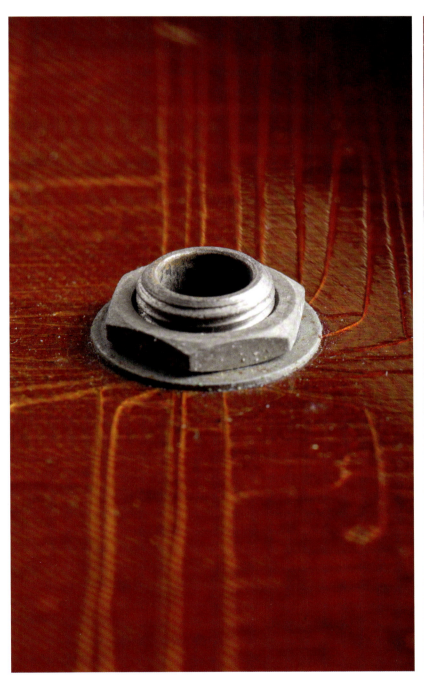

MARTIN D-28/12
Natural
1976
#381839

I've always loved the sound of 12-string acoustic guitars, so I started using one as soon as I could. People often make a big deal of how hard they are to play or keep in tune, but they're worth the effort. It's the sound of 'Get The Message' by Electronic, and 'Half A Person' and 'Unhappy Birthday' by The Smiths. ∎

INTERVIEW: PART 2
Johnny Marr with Martin Kelly

JM When I was writing the early Smiths material, I had an idea – a concept, really – because I was listening to a lot of sixties girl group music: The Shirelles, The Marvelettes, and particularly The Shangri-Las – I loved those records. I fell into that because Patti Smith liked the girl groups and The New York Dolls liked the girl groups. I was always into buying old records and I was super into Motown, so I was thinking that if I could take what those records were doing and do it on guitar, through a Fender Twin Reverb, what would that be like? And as a concept, it was pretty cool. Obviously, it came out as my own thing because there's a British thing that comes out of it as well. I was completely unaware of what my own technique was. I didn't think, 'Oh I'm quite good at arpeggios.' I just was good at arpeggios, but I didn't know that until people started telling me. I didn't think, 'I'm a rhythmical player who plays all of those melodies inside of that.' I just know that now. At seventeen or eighteen, all I knew was what I wasn't, and where I wanted to get to. I just wanted to write really good songs. Instrumental songs, that's what I was focused on in the beginning.

I'd put a chord in a different place on the neck or change its shape around and would make a discovery. I noticed that certain chords sounded more like how I felt, like I was playing something that was personal to me and that I could relate to. I was looking for things that evoked a sense of yearning, but with a kind of optimism, and that started to develop into an identity of my own. Bit by bit, these discoveries became my vocabulary on the guitar, and I was sounding more and more like myself.

It was only when I got well known that people started saying, 'You are this' and 'You are that' or 'You are chimey' and 'You are jangly'. I didn't know that at the time. In fact, it was only when The Smiths walked off stage after our very first gig on 4 October 1982 that my friend Joe Moss, who became our manager, said, 'You're really different; I've never heard anyone play like that.' I genuinely didn't know this. I wanted to play like a pop Bert Jansch or someone, but I'd discovered Nile Rodgers by then and I was wigging out over the Stooges' *Raw Power*, too. And on those early Smiths songs, I was also really trying to play like Motown. So, there's all kinds of different influences.

It always surprises people when I mention James Williamson and Nile Rodgers as influences, because they think I'm nothing like them. But if you take a lot of the distortion off, I actually am. As Iggy Pop once said, 'James fills up every corner,' which is sort of my approach, too. My sister introduced me to Chic and Nile Rodgers, which was a really important discovery for me. It was what Nile was doing with his left hand, those really beautiful chord progressions, that really clicked with me. So, there's a lot of that in early Smiths records, songs like 'This Night Has Opened My Eyes', 'I Don't Owe You Anything', 'The Boy With The Thorn In His Side' and 'Hand In Glove'.

I was on a mission of discovery, back at my parents' house, plus I was desperate to get a band going. I felt like I'd sort of done my apprenticeship really – but I hadn't, I'd only half done it. Then, when I moved out, when I formed The Smiths, I moved into an attic in a family's house a couple of miles away. When The Smiths formed, I knew straight away that musically it was going to work.

Just before forming the band, I traded in my black Les Paul for a Gretsch Super Axe, because I wanted a Gretsch and that model was the one I could afford. I wanted something a bit more fifties, because an aspect of being a guitar player and a guitar obsessive is a guitar's associations. Also, when you're young you want to be modern. Before the 1980s, the Les Paul had influenced a lot of punk players because of three people: Mick Ronson first and foremost, Pete Townshend and Paul Kossoff. They had all been associated with the Les Paul, but by 1979, 1980, 1981 when I was starting out, there was a new wave coming. A lot of guitarists of my age were starting to get interested in Gretsches and 1950s guitars – I'm thinking of Roddy Frame in Aztec Camera and Billy Duffy with Theatre Of Hate, Matthew Ashman of Bow Wow Wow and Edwyn Collins with Orange Juice. So, it was almost like the Gibson Les Paul and Fender Stratocaster fell out of favour after the 1970s, and guitarists my age wanted something different. John McGeoch, Bill Nelson, Andy Partridge and Stuart

Adamson were all playing Yamaha SG-100s, a new guitar that I disliked at the time, but I've got a bunch of them now – amazing instruments. At the time though, the Gretsch was right for what I wanted to do, and I wrote 'This Charming Man' and the first set of Smiths songs on the Super Axe.

The next guitar I got, which is really significant, was the black Rickenbacker 330. I bought it new from A1 Repairs with the money from my first publishing advance. I loved the idea of a Ricky, but what really made me want to buy it was that I knew it would make me play in a different way, and it would certainly make me write in a way I should write. That's a thread that recurred 30 years later, when I eventually found the Fender Jaguar. It was exactly the same consideration: this guitar will stop me doing things I don't want to do. I think I was quite self-aware of that really.

A lot of people think the opening riff to 'This Charming Man' was played on a 12-string, but it was actually the black Rickenbacker double-tracked with a 1954 Telecaster. With hindsight, I can see that it's quite an unusual guitar part, but when people were making a big deal of it over the years, I'd just be like, 'Oh well', because I did it. But now I can see that it is quite unusual. I didn't have any time to think about it when I came up with it. That's the thing – I can tell you what it wasn't. So, it wasn't that I had this riff, which can happen, where I'm walking around and I'm sitting on the train or the bus and I think, 'Oh, how would I play that?' It wasn't that. It wasn't that I thought about musical theory and went, 'That's quite unusual, that's quite clever.' It wasn't that at all. It was just that I played these chords, the underlying chords, into my TEAC and I went G G G G G G G D, A minor, C D G. That was the bed of 'This Charming Man', and I pressed stop, and then I rewound it. I went to do the overdub and I just did that riff, straight away, on top of it. My fingers just went to that and I did it really casually. I didn't think, 'That's an amazing riff.' Then the other bit in the chorus is also really unusual, but I didn't think about that either. It just came out of the sky really. So, it was a really lucky one. The whole song was actually written like homework I had to do, because we had a John Peel session coming up and I thought the band needed a jolly one. I'm in that rare club of people who have songs that people want to talk about forty years later. It's an amazing thing.

•

JM The Smiths moved to London at the start of 1984 and I was living in Earl's Court. I bought a Gibson J-160E in Andy's on Denmark Street and an Epiphone Casino from Earl's Court Guitars, which was conveniently and also dangerously close to where I lived.

Occasionally, I'd meet one of the older guys in one of the shops. I remember Gary Moore was in there one time, and he was great – a really nice guy. I was amazed that someone like him would know my stuff. That happened a few times, with Rory Gallagher and Ronnie Wood. Once, I got a message from Pete Townshend saying he liked my playing. It was like I'd been accepted into a fraternity, which was a massive boost and also kind of humbling. I find that most guitar players are generally very welcoming, encouraging and just really nice.

Around this time, when The Smiths started to have a run of hits – 1985, I guess – I became friends with a guy called Alan Rogan, who is sadly no longer with us. He was the guitar tech for Pete Townshend, and he'd also looked after Keith Richards, Eric Clapton and George Harrison among many other people over the years. Alan was one of the most respected guitar experts there'll ever be, and because he was working for The Who I got a few guitars from John Entwistle. Alan also found me my three pickup Les Paul Custom in Minneapolis, after calling me saying, 'Listen, you're going to love this guitar, I think you should get it.' Sometimes I'd say, 'I really fancy an early Strat, what do you reckon?' and he'd say, 'I know where there is one.' I was using a lot of different guitars in The Smiths, probably compared to my peers. I was making records with more layered guitar sounds, which is what I'd started off learning to do back in my bedroom at my parents' house.

Keith Richards was a really important figure for me, and I first got to meet him in 1986. He was really good to me, and I actually found him to be very cool and quite shy even. Over the years, I've always had the same experiences with the musicians from that period: that they just want to play. Keith Richards, Paul McCartney, Nile Rodgers … they might have some good stories or whatever, but the conversations always come back to playing. And if there are instruments there, that's what happens. To me that's inspiring. It's not sitting around and conceptualizing; it's like, 'Let's play.' I had a few great sessions with Keith – which was mind blowing for me as a twenty-three-year-old guy – and that was fantastic. Paul McCartney – he'll wear you out; he just wants to play and play and play.

MK You left The Smiths in June 1987 after you finished the final LP *Strangeways, Here We Come*. Within weeks of the record's release, you were approached to join The Pretenders.

JM The Pretenders had been supporting U2, who were promoting *The Joshua Tree* with a huge tour of North America. There was a short break before the last leg, when their guitar player, Robbie McIntosh, left the band. Someone from The Pretenders' office sent me a message saying, 'The Pretenders want to continue with the U2 tour and they need someone who can stand in. How about it?' The first show was about a week after I got the call. I was living in South Kensington and was dealing with the end of The Smiths. The idea of escaping to the USA to play guitar with The Pretenders sounded like a perfect solution at the time, so I said yes. The amazing thing was that I had three days to learn The Pretenders' set before the first rehearsal, and we were only going to rehearse three times. I stayed up all night for two nights, rewinding and fast forwarding a video of them playing the set. It was an amazing opportunity and a challenge, and just one of those things that happened in my life that was both pretty abstract and quite a big deal when I look back on it. Something I grabbed because it felt, at the time, like it was happening for a reason. I didn't know whether I was going to join the band for an album, or if there was a long-term plan. It was just, 'Can you learn a Pretenders set to go out on tour with U2?' Some of the shows were in front of 100,000 people. I just had to handle it.

So, first I had to walk into a rehearsal room with a bunch of strangers. I think I was twenty-three at the time and the band were much older than me. I wasn't sure they were even aware of what I was about. All they knew was that this kid was the new guy and we had a major tour in a few days. So it was a bit nervy. I loved The Pretenders' first album, which I knew by memory from learning it when it came out, things like 'Tattooed Love Boys' and 'Kid', and I knew 'Talk Of The Town' and stuff like that. Their original guitarist, James Honeyman-Scott, had been a big inspiration – he was brilliant. So, it was fucking weird, but one of those times when I thought this just might be great, and it turned out to have quite a long-lasting effect on my life. I didn't realize it until much later, but so much of my experience fronting a band I learnt from Chrissie Hynde. Loads of things. The way she handles fame and attention. Some little things, like the way you walk on stage.

A way of being in rehearsals. It's been pointed out to me that some of the things I do onstage are her moves, because I'm forever taking my hand off the guitar and pointing, which I think comes indirectly from Ray Davis via Chrissie. She really did more for me than she will ever know. She was like a big sister and had a massive influence on me.

MK After the U2 tour, you recorded two songs with The Pretenders for the soundtrack of the film *1969*. Then, at the start of 1988, you received another invitation, this time from your old friend Matt Johnson of The The, who you'd known prior to forming The Smiths. You became a permanent member of The The from 1988 to 1994 and played guitar and harmonica on those seminal albums *Mind Bomb* and *Dusk*, as well as touring extensively with the band. It seems like you and Matt had a very strong personal and musical bond.

JM When I joined The The, my role was a little bit different because the band had quite a lot of keyboard parts, plus Matt was also playing guitar with his own unique style. I had to try to employ enough technique and imagination to enhance the songs and to try and add a slight element of surprise. For instance, there's a song on *Mind Bomb* called 'Kingdom Of Rain'. The song would have held up without my part, but when the chorus kicks in with Matt and Sinead O'Connor, my guitar is unexpected and it elevates the song. That was me doing my job. A song like 'Slow Emotion Replay' has the arpeggio hook, which brought a sort of joy and an uplifting vibe to an already lovely track. When I came up with the part, I was playing my feelings in the moment. That's what's so great about music. I'm not sitting back thinking, 'What would be clever?' It was a very happy time and I loved the song. When I recorded the harmonica part, I was so happy hearing the playback that I started singing down the harmonica mic – which I had no business doing, but it stayed on the record.

Playing in The The was a pioneering time for me as a guitar player. I didn't feel like I was any less involved in The The than I was in The Smiths, believe it or not. It was just that my role was different, and I welcomed that role because it actually made me a better guitar player. And when I'm in a band, I'm all in. Anyone who's been in a band with me will tell you that. The The was the band that I would have been in around 1982 had I not started The Smiths. And because Matt Johnson and I knew each other so well, I knew what to do in the band.

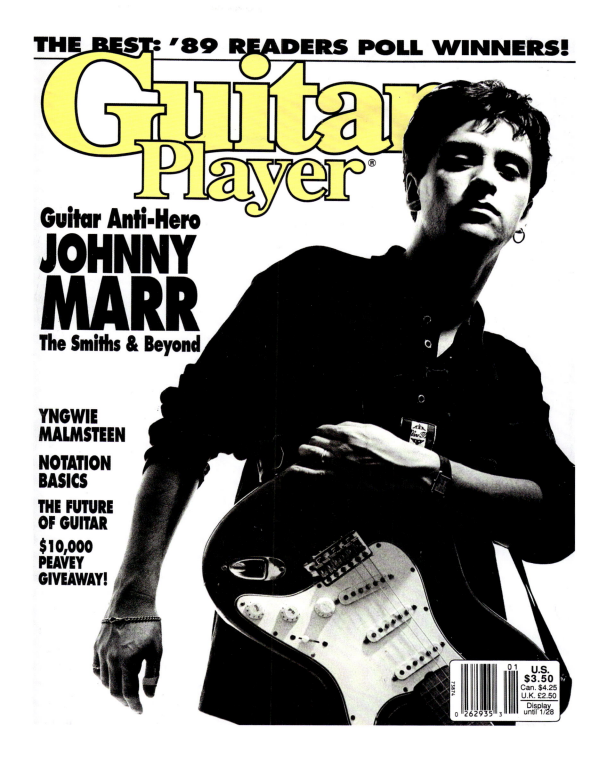

FENDER STRATOCASTER
Lake Placid Blue
1964
#L18068

Another *Strangeways...* guitar. Later, I loaned it to Alex Kapranos of Franz Ferdinand for *Right Thoughts, Right Words, Right Action*. ∎

FENDER TELECASTER
Black
1967
#205962

I wanted a Telecaster that I could play lead on, so I swapped out the
pickups on this for a more powerful sound. This guitar was used a lot with
Electronic and with Billy Bragg on 'Sexuality' and 'Cindy Of A Thousand
Lives'. I played it for my set at Patti Smith's *Meltdown* concert in 2005 at
the Royal Festival Hall in London. After the show, Patti signed the back
of it and put in quotes: 'People have the power ... Patti Smith'. Her message
wore off over the years, and at one point it just said 'the Smith'. ∎

GIBSON ES-335/12S
Sunburst / Cherry / Black
1968 / 1967 / 1965
#905064 / #878208 / #346460

The set of three 335 12-strings that I acquired in the late 1980s. The red one was used by Chrissie Hynde for the song 'Back On The Chain Gang' when I was in The Pretenders and we opened for U2 on *The Joshua Tree* tour. I also played it on The Pretenders song 'Windows Of The World'. ∎

NATIONAL STYLE O RESONATOR
Nickel-Plated Steel
1936/37
NO SERIAL NUMBER

Bought in London, Ontario, on *The Queen Is Dead* tour, this is a rare style of National as they are usually decorated with a palm tree motif. I liked it without that extra detail. ∎

FENDER STRATOCASTER
Petrol Blue
1983
#E301671

A one-off colour Strat, given to me by Fender in 1988. It became my main onstage guitar with The The and Electronic and was the start of my relationship with the Fender guitar company as an endorsee artist. ∎

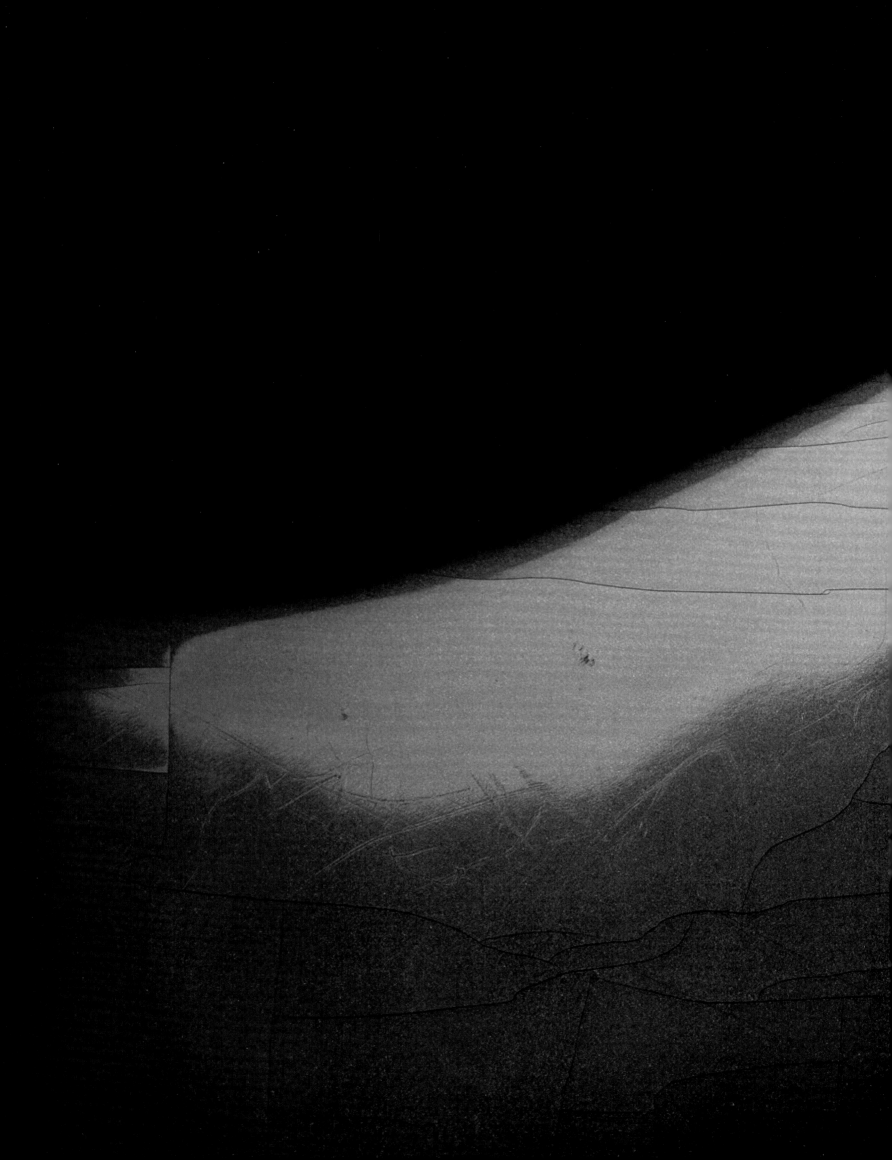

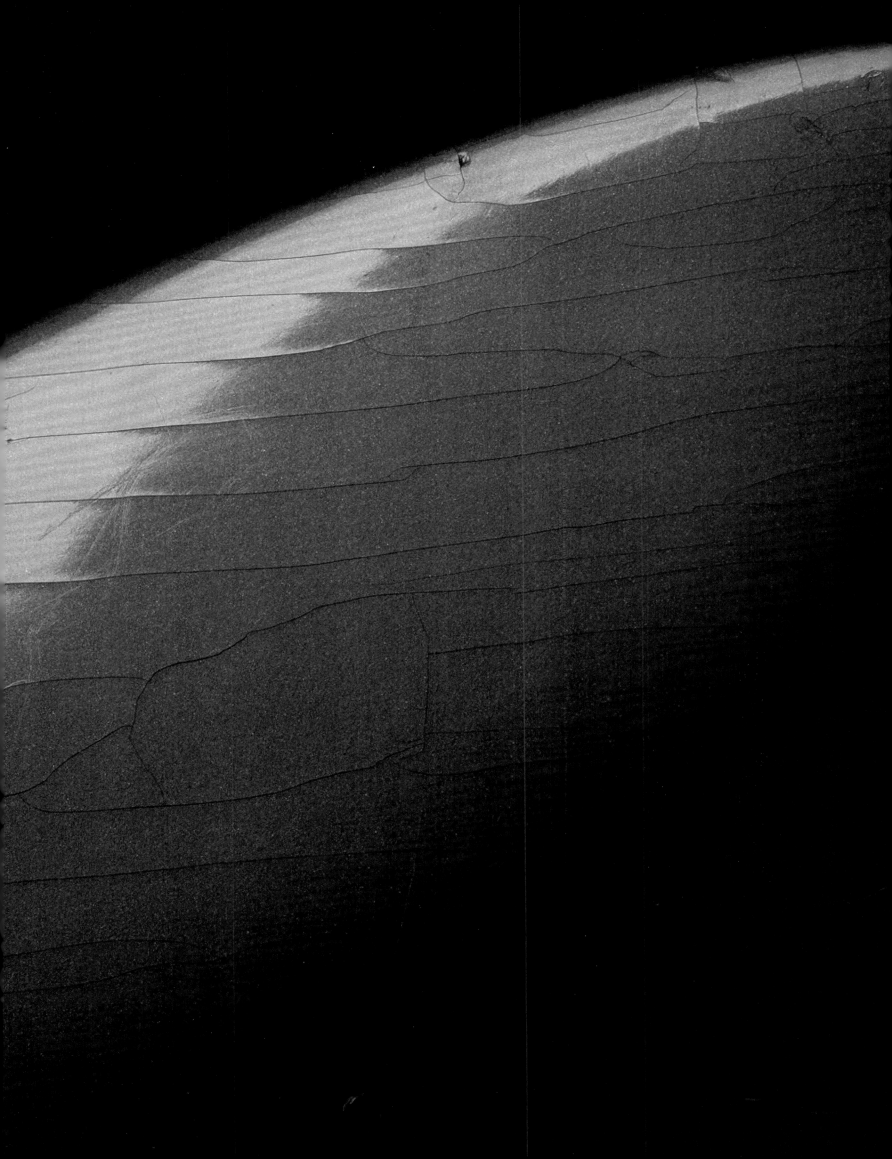

GIBSON LES PAUL CUSTOM
Black
1960
#00825

GIBSON LES PAUL SPECIAL
TV Yellow
1960
#02049

GIBSON LES PAUL GOLD TOP
Gold
1957
#70331

When Radiohead were recording the album *In Rainbows*, I loaned them this guitar and a few others.

'It's great having a friend like Johnny because you can talk about the universe but you can also talk about Fender Deluxe Reverbs and Gibson 335s – everything as important as the rest. Radiohead were embarking on making In Rainbows *in 2006 and I phoned Johnny and said, "I think we've got a good album here". Johnny said, "Do you want to borrow any of my guitars?" and at that stage our guitars were really quite limited; Thom didn't have any and Jonny (Greenwood) had been getting into modular synths, and I think Johnny (Marr) said, "I have an idea of what I'll send down". He had his tech deliver the 1957 Les Paul Gold Top,*

a 1964 SG and a 1964 white Fender Stratocaster and he said, "This is 'The Boy With The Thorn In His Side' Strat". We had really no idea about vintage guitars. I was trying this 1957 Les Paul thinking, "Oh my god, this is the most extraordinary guitar I've ever played – I need to get one of these." There were two sessions for In Rainbows, *and these three guitars were part of the arsenal for the sessions. "Bodysnatchers" was one of the first songs we recorded, with the five of us playing live, pretty gnarly, like a nest of angry guitars, and I'm pretty certain I played the Les Paul on that. I'd never played a guitar like the Les Paul – there's something about the neck on that, the classic Gibson 1957 neck, this beautiful solid piece of wood and the age and tone and resonance, and it looked incredible. I know that during the session Thom really got into using Johnny's SG. He fell in love with it, and I think he played it on "Reckoner" and "House Of Cards". Basically, the SG became his go-to guitar on that record, and he may have played it on "Weird Fishes" as well. At the second sessions, we were all suddenly like: Oh my god, these vintage guitars have an incredible sound to them – not only have they got the provenance of Johnny and what he played on them, but in themselves they were amazing guitars. I remember I went to a guitar store in Bath and I ended up buying a 1964 Epiphone Casino that became part of my arsenal. And so, in a way, Johnny lending us these vintage guitars opened the door to us appreciating the sound and understanding the vintage guitar. How great is that – we are all huge Johnny Marr fans, and I felt he was sort of like a big brother watching over me. I remember we finished mixing it, and I drove up that night from Oxfordshire to Manchester because I was so proud of this record, and I wanted him to hear it because Johnny is one of the most supportive people you will ever meet in life and was so supportive of Radiohead. When you have the blessing of Johnny, it means a lot. You get Johnny Marr giving you the thumbs up, saying you guys are great. You know it's a big deal lending someone your guitars; there's a lot of trust. You can't quantify how important those gestures are, but they actually mean a lot, and I don't think it's any coincidence that him lending us those guitars was part of making, for me, what is one of my favourite Radiohead albums – it's one I really love.'*

ED O'BRIEN

GIBSON ES-350TDN
Natural
1962
#52168

GIBSON S-1
Sunburst
1976
#00177533

FENDER STRATOCASTER
Black
1978
#S892232

This is the guitar I got when I took a trip to a guitar shop in Doncaster with
Noel Gallagher, shortly after we first met in 1992. It has nine pickups and
eighteen switches and is a guitar tech's nightmare. I have no idea who built
it ... a madman. When I saw it in the shop, I thought: 'If Kraftwerk played
guitars ... they would be playing this.' It finally came into its own on 'Spirit

GIBSON SG STANDARDS
Cherry
1964 / 1964
#242468 / #240814

When I formed my first solo band, The Healers in 1999, I was inspired to get an SG because I loved the sound of Mick Taylor when he was in John Mayall's band, Bluesbreakers, and also because of Pete Townshend in the late 1960s and early 1970s.

I was obsessed with the SG on the left in a way I had never been about a guitar before, and it was the only guitar I played during that time. The Healers had been on an extensive tour across the USA, Japan and Europe, and we ended in London playing the last night at the Scala Theatre. As soon as the show ended, I was back in my dressing room as some of my friends started to arrive, when my guitar tech ran up to me in a total panic saying the SG had been stolen. Everyone was in shock when we realized it had gone. What happened was that, when the band had walked off the stage, a guy had just climbed up out of the audience, run over to my guitar, grabbed it and walked out the front door. No one could believe it had happened. That particular guitar had been so important to me. I had medals and coins and

a few bits of jewellery hanging off it as souvenirs. It was my main guitar and couldn't be replaced. The next day, people were calling me from guitar shops to say how sorry they were that it had been stolen and that they'd be on the lookout for it. I put a message out on the radio with a reward for any information that might lead to it being returned, but I didn't hear anything.

Ten years later, when I was in Toronto on tour with The Cribs, my manager called me to say that a detective from King's Cross had contacted him because he was a fan and was interested in reopening the case. He suspected that the guitar might still be in the area. I was dubious about him finding it, but I filled out another report and left it at that. A couple of weeks later, the detective contacted us again to say that he'd discovered who had stolen my guitar. He tracked it down and retrieved it. It was amazing. After all those years, I got the guitar back. Some people will have heard that this guitar is called 'Betsy'. The reason it's called Betsy is a bit convoluted, but is essentially me trying to be ironic. I've never felt the need to give any of my guitars names, but after seeing the TV show *King of the Hill*, where the great Hank Hill names his guitar Betsy, I thought it would be amusing to give my guitar the same name. That's my story and I'm sticking to it. The SG on the right was bought as a replacement for the stolen one, and it became one of my favourite guitars. It was later used by Thom Yorke on the Radiohead album *In Rainbows*. ∎

GIBSON SG STANDARD
Natural
c. 2000
NO SERIAL NUMBER. CUSTOM-MADE ONE-OFF

Made for me by master builder Roger Giffin at Gibson. ∎

YAMAHA LLX-400
Natural
2001
#XL458J

MARTIN D-35 [OVERLEAF]
Natural
1969
#247718

One of the biggest influences on my guitar playing was Bert Jansch. I was introduced to his music in my teens, and he remained a touchstone for me and an ongoing source of inspiration, as he was for so many guitarists, such as Jimmy Page, Donovan and Neil Young – as well as many other guitarists who don't even realize they've been influenced by him. I got to know Bert around 2002. We became friends and started to play together onstage and in the studio. I always used my Martin D-35 when we worked together. It seemed appropriate as Bert was one of the reasons why I played a Martin acoustic in the first place. When he passed away in 2011, his family very kindly gave me his Yamaha LLX-400. It's an honour to have it. ∎

RICKENBACKER 370/12
Jetglo
1989
#I26944

There have been a few occasions when I've wandered into a guitar shop and just known that something special was going to happen. I got this guitar in Los Angeles in 2000. I'd been without a guitar for a few days and was really missing playing, so I went into the Guitar Center on Sunset Boulevard and immediately spotted this 12-string Roger McGuinn prototype among the hundreds of other guitars in the shop. I usually try to be pretty low-key when I'm in guitar shops, but as soon as I got this guitar in my hands, I had to have it. I plugged it in and turned it up really loud and started blasting for about thirty minutes. A memorable morning on Sunset. I used this guitar extensively with The Healers, Modest Mouse and onstage with R.E.M. for the songs 'Fall On Me' and 'Man On The Moon'. ∎

GRETSCH 6120 DOUBLE-CUTAWAY
Western Orange
1964
#79389

I played this guitar with Neil Finn in Auckland in 2000, when we formed a band called 7 Worlds Collide with our friends Eddie Vedder, Ed O'Brien, Lisa Germano, Sebastian Steinberg and Philip Selway. I had the f-holes cut out instead of them being painted on, which is how they usually are. ∎

**FENDER TELECASTER
CUSTOM SHOP RELIC**
Blonde
2004
#R17662

Given to me by Fender in 2004. ∎

Happy Birthday Fender®

"Thanks for the glorious noise!"

Fender
60
DIAMOND ANNIVERSARY
1946 - 2006

© 2006 FMIC   JOHNNY MARR photo by Mick Rock

FENDER TELECASTER
Black
1976
#7624968

I used this guitar – bought in Louisville, Kentucky, birthplace of Muhammad Ali – on tour with Modest Mouse. It's occasionally known as the 'Louisville Slugger'. ∎

FENDER JAZZMASTER
Sunburst
1965
#L90136

Bought in Tupelo, Mississippi, one day in 2007 when I went to visit the childhood home of Elvis Presley, and used extensively on the Modest Mouse album *We Were Dead Before The Ship Even Sank*. To non-guitar experts, the Jazzmaster and Jaguar appear to be very similar. The differences are that the Jazzmaster is physically bigger with a longer scale length and wider pickups that give it a slightly 'bigger' sound, whereas the Jaguar has a more 'focused' sound, operating in a slightly different area sonically. ∎

FENDER JAZZMASTER CUSTOM
Natural Wood
Early 2000s
#312130

Made for me by Dennis Galuszka at the Fender Custom Shop. I wanted something that had the body and neck of a Jazzmaster with the wiring of a Jaguar. I added the Gretsch pickups later. ■

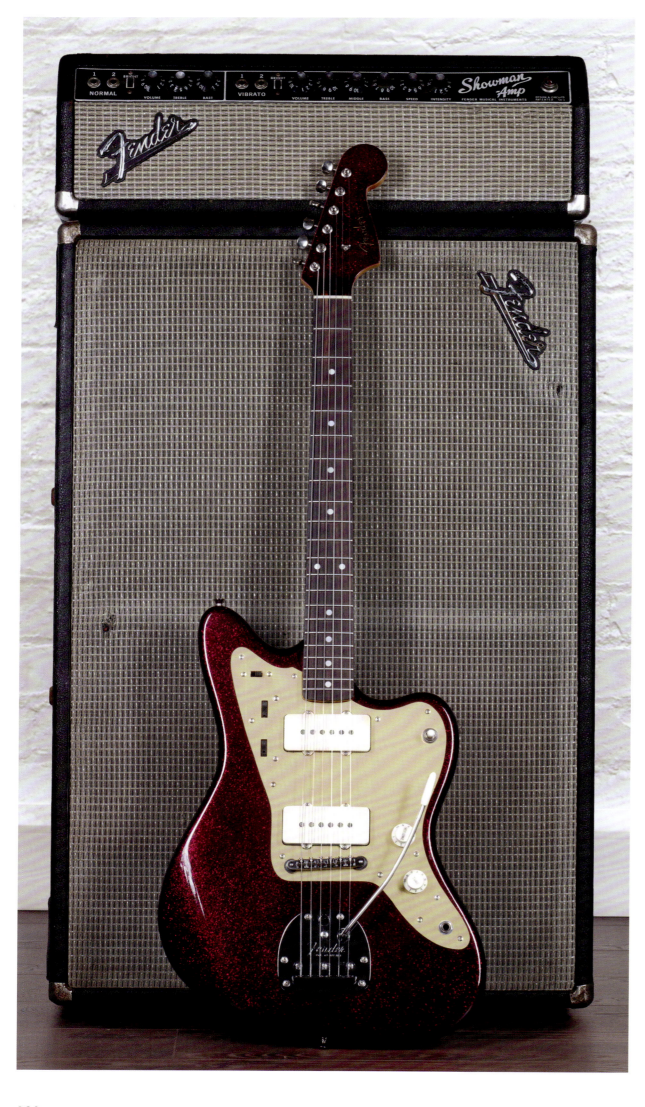

**FENDER JAZZMASTER –
J MASCIS SIGNATURE
MODEL**
Purple Sparkle
2007
#050471

Given to me by J Mascis
after I played with him and
Dinosaur Jr. in New York
in 2012. ∎

**FENDER
JAZZMASTER**
Inca Silver
1962
#76840

FENDER JAZZMASTER
Shoreline Gold
1965
#L82207

FENDER JAZZMASTER
Olympic White
1996

190

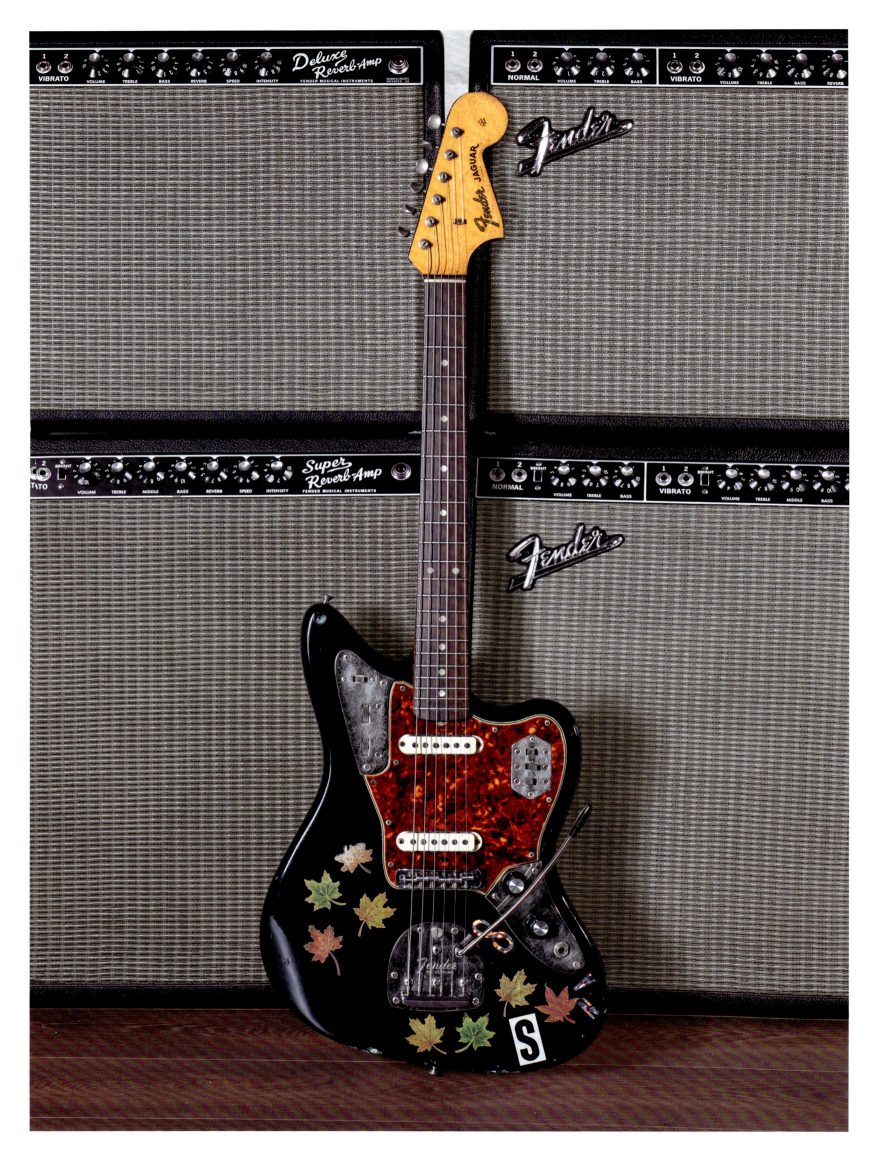

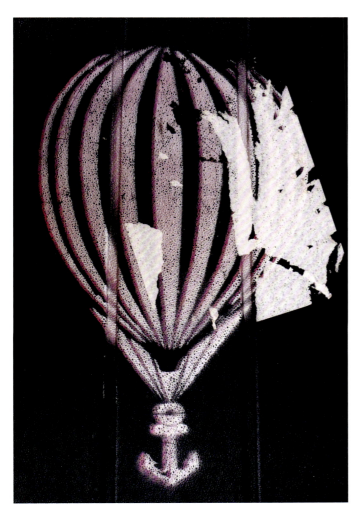

FENDER JAGUAR
Black
1965
#L95414

In 2005, Isaac Brock called me to ask whether I'd be interested in writing with Modest Mouse. I was a fan of the band and I thought it might be an interesting musical experiment. I went out to Portland, Oregon, where the band was based, and the night I arrived, Isaac and I got together in his attic to play some guitar and to start to kick around ideas. I had brought over my fairly new Custom Shop Telecaster, which I played through my usual Fender Deluxe amp. Isaac played a custom-made Wicks guitar through a Fender Super Six amp, which is possibly the loudest amp in the world, and I realized my Telecaster couldn't compete with the enormity of Isaac's sound. I'd noticed this black Fender Jaguar in the room an hour or two earlier, and without thinking about it, I picked it up and started to play. Immediately, something great happened and I thought: 'Where have you been all my life?' At the same time, Isaac – who had been working his way through an impressive jug of wine – got into my face and scowled: 'Got any riffs?' It just so happened that I did have a riff that I'd been playing around with for a week or so, and then something about this black Jaguar made me play the same riff in a new kind of way. It was a magic moment,

and one that doesn't happen too often, where inspiration comes down from the sky like lightning and strikes two people at once. Isaac started improvising the entire lyric: 'Well, it would've been, could've been worse than you would ever know. The dashboard melted, but we still have the radio.' The two of us immediately started writing the song 'Dashboard', which would become the first single on our next album. Once we'd got most of the song done, Isaac asked me, 'Got any more riffs?' and it just so happened that I did have another riff: it became the song 'We've Got Everything'. I knew that discovering this particular guitar was a big moment, and so I asked Isaac if he'd sell it to me. Luckily, he said he would. It was the start of my journey as a Fender Jaguar player. Put simply, the guitar made me play exactly like myself, and in a direction that I felt was exactly where I needed to go. I moved to Portland to join the band full time, and we embarked on an inspired few months during which we wrote *We Were Dead Before The Ship Even Sank*. The album went to Number One in the USA, and 'Dashboard' won the award for the most played alternative track on US radio in 2007. The band toured extensively for the next few years and, as I got into using the Jaguar more, I started to wonder if I could improve the design and bring it more up to date. These improvements were technical things, like changing switches and replacing the bridge as well as designing a new type of tremolo arm that would remain in a fixed position. Eventually, I would replace the neck and pickups as I worked obsessively on all aspects of the guitar with my luthier friend Bill Puplett, who has worked on all my guitars since 1988. The Fender company heard that I was working on building a custom-made Jaguar and invited me to make my own signature version. To have my name on a headstock and be in the company of people like Chet Atkins and Les Paul is a real honour. When I got the prototype guitar to what I thought was near perfection, I handed it over to Fender and they manufactured it to my exact specifications, which means that every one of them in any shop in the world is exactly the same as mine. The signature Jaguar has been a huge success, winning the Instrument of the Year award at the Frankfurt Music Convention for an unprecedented two years running. The making of my own signature guitar is one of the things I'm most proud of in my career. I've never felt the need to play any other type of guitar onstage, and I couldn't and wouldn't change a thing about it. ∎

FENDER JAGUAR
Olympic White
1962
#L19145

I played this white one, as well as the black one, in Modest Mouse before
I started building my signature model. ■

FENDER JAGUARS
Olympic White
2008 / 2009 / 2008
PROTOTYPE / #V202051 / PROTOTYPE

In the photo opposite, the first signature prototype is on the left and the second is on the right. Both were built during 2008. In the middle is the first 'main one' that I used from 2009, when I joined The Cribs and throughout my first solo tours. The artwork on the back is ink and was done in Australia by my friends Aly Stevenson and Ory Englander, who created the Johnny Marr fan site and fanzine *Dynamic*. The images are of Aldous Huxley, Marcel Duchamp's *Bicycle Wheel* and the Chrysler Building.

Jazzmasters and Jaguars fell out of favour in the late 1960s, but were brought back into vogue by the New York band Television in the late 1970s. At that point, they were cheaper than the more desirable Fender Stratocasters and Telecasters. They were taken up again in the 1990s, most notably by Sonic Youth who, in turn, inspired bands like Dinosaur Jr. and Nirvana. These types of Fender guitars are sometimes known as 'offset' guitars because of their unique angular shape. ■

FENDER JAGUAR –
JOHNNY MARR SIGNATURE
Metallic KO
2012
#V204834

My signature model has been made in a series of custom colours.
This was the first one. It was a brand-new colour that we called 'Metallic
KO'. The idea came from seeing an old 1965 Jaguar that had corroded
over the years and turned into this shade of burnt orange. ∎

FENDER JAGUAR –
JOHNNY MARR SIGNATURE
Sherwood Green
2013
#V1316595

One of the most popular of the custom colour series.
Now very collectible. ∎

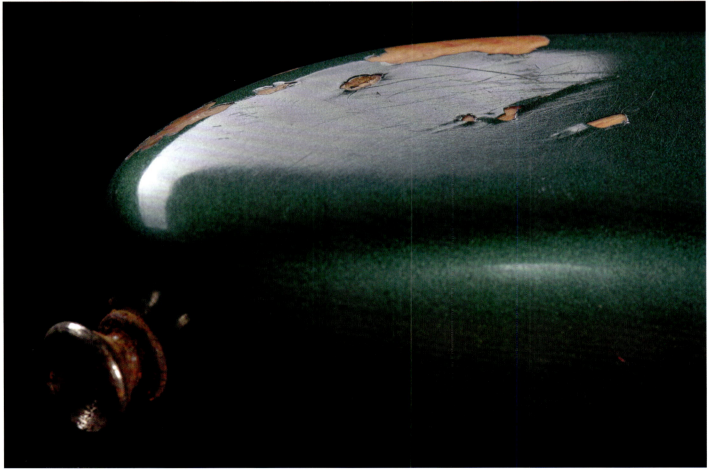

**FENDER JAGUAR –
JOHNNY MARR SIGNATURE**
Candy Apple Red
2021
#V2096168

FENDER JAGUAR –
JOHNNY MARR SIGNATURE
Black on Black
2012
#V202055

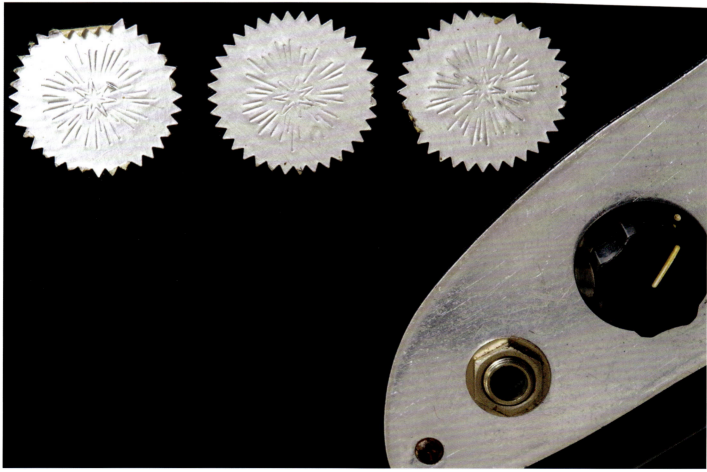

FENDER JAGUAR –
JOHNNY MARR SIGNATURE
Lake Placid Blue
2017
#V212043

This guitar is fitted with the sustainer pickup system from the Ed O'Brien Signature Fender Stratocaster. It gives the effect of all six strings sustaining at once, creating an orchestral or ambient type of sound. ∎

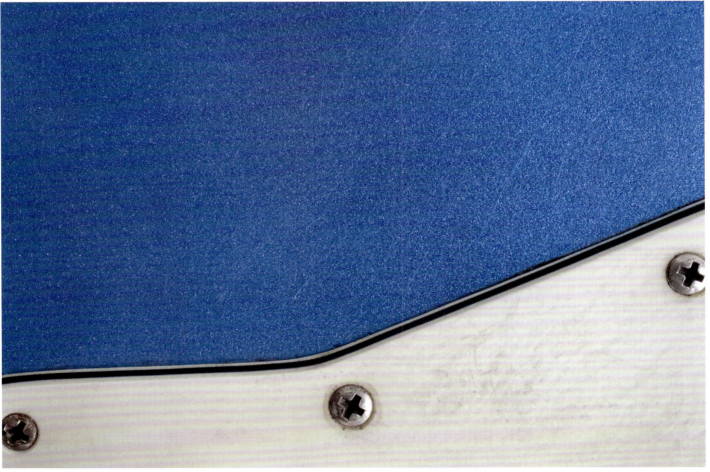

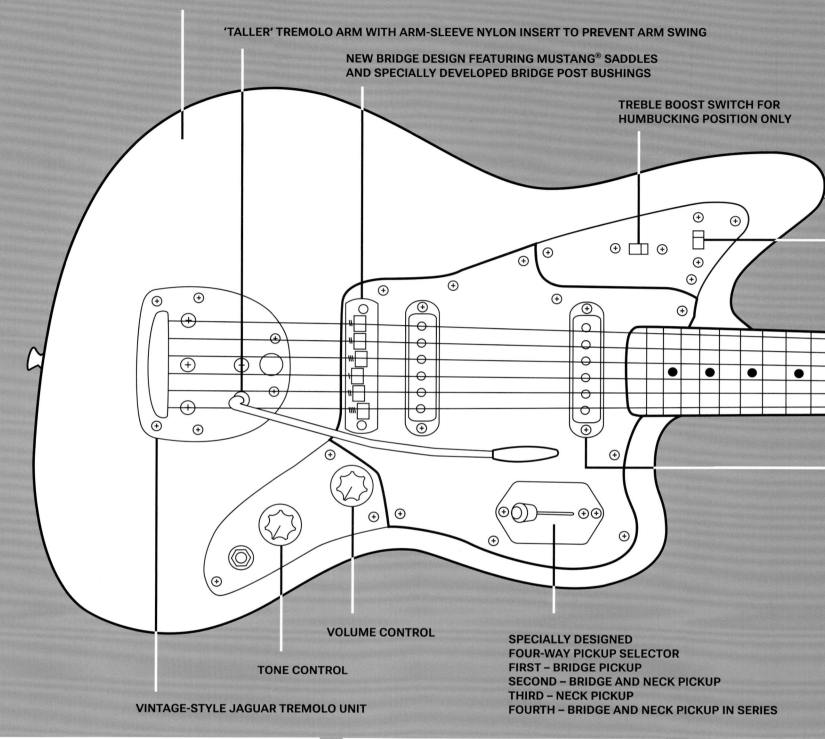

AVAILABLE IN OLYMPIC WHITE AND METALLIC KO FINISHES

'TALLER' TREMOLO ARM WITH ARM-SLEEVE NYLON INSERT TO PREVENT ARM SWING

NEW BRIDGE DESIGN FEATURING MUSTANG® SADDLES
AND SPECIALLY DEVELOPED BRIDGE POST BUSHINGS

TREBLE BOOST SWITCH FOR
HUMBUCKING POSITION ONLY

VOLUME CONTROL

TONE CONTROL

VINTAGE-STYLE JAGUAR TREMOLO UNIT

SPECIALLY DESIGNED
FOUR-WAY PICKUP SELECTOR
FIRST – BRIDGE PICKUP
SECOND – BRIDGE AND NECK PICKUP
THIRD – NECK PICKUP
FOURTH – BRIDGE AND NECK PICKUP IN SERIES

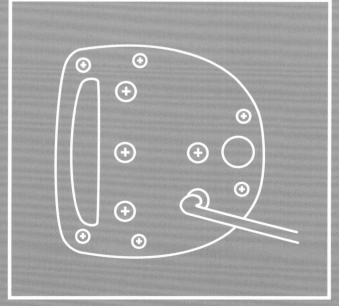

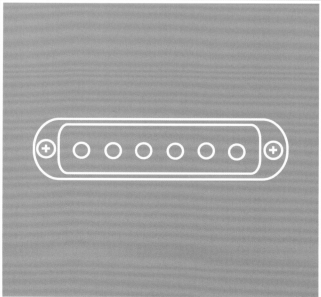

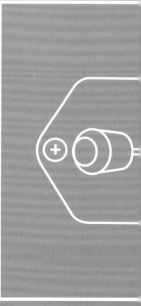

JOHNNY MARR JAGUAR® SIGNATURE MODEL

HIGH PASS FILTER FOR ALL PICKUP CONFIGURATIONS

VINTAGE-STYLE KLUSON TUNERS

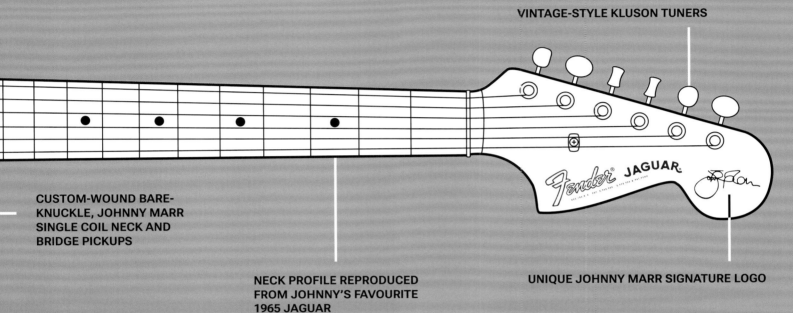

CUSTOM-WOUND BARE-KNUCKLE, JOHNNY MARR SINGLE COIL NECK AND BRIDGE PICKUPS

NECK PROFILE REPRODUCED FROM JOHNNY'S FAVOURITE 1965 JAGUAR

UNIQUE JOHNNY MARR SIGNATURE LOGO

THE JOHNNY MARR SIGNATURE JAGUAR IS A FANTASTICALLY NON-STANDARD VERSION OF THE MODEL THAT IS AS DISTINCTIVE AS THE SOUNDS MARR WRINGS FROM IT

OTHER PREMIUM FEATURES INCLUDE THE CLASSIC JAGUAR 24" SCALE LENGTH LACQUER-FINISHED ALDER BODY, 7.25" RADIUS ROSEWOOD FINGERBOARD WITH 22 VINTAGE-STYLE FRETS, MASTER VOLUME AND TONE CONTROLS, THREE-PLY PICKGUARD AND CHROME HARDWARE, ACCESSORIES INCLUDE A CUSTOM CASE WITH BLUE CRUSHED VELVET INTERIOR, STRAP, CABLE AND FLATWOUND STRINGS

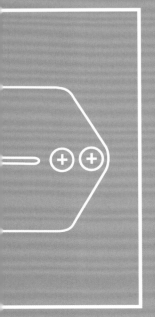

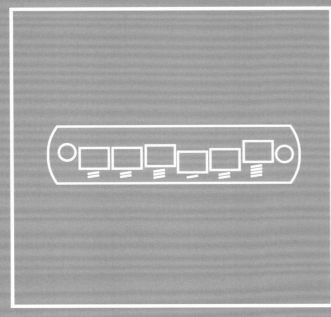

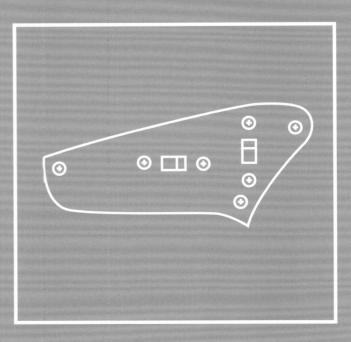

FENDER STRATOCASTER
Black
1963
#69320

Bought from my friend Billy Duffy of The Cult, I used this guitar on tour with The Healers. ∎

GIBSON J-200
Sunburst
1968
#896645

INTERVIEW: PART 3
Johnny Marr with Martin Kelly

MK In late 1987, you formed the supergroup Electronic with Bernard Sumner from New Order and Neil Tennant of the Pet Shop Boys and released the debut single 'Getting Away With It', which was a huge hit in the USA. That band released three albums throughout the 1990s, and on the second one, *Raise The Pressure,* you were joined by Karl Bartos from Kraftwerk. How did Electronic come to be?

JM Electronic happened after Bernard and I met up in San Francisco in September 1987, on the last night of a New Order tour. He'd invited me to the show and we spent the night hanging out. He then invited me to write some songs with him. I went from there to New York to work on a Dennis Hopper movie called *Colors*, and then went back to Manchester where Bernard and I started writing. We'd get together in my home studio a few nights a week every week, and nine years later we were still writing together. I was keen on working with synthesizers and technology and was getting very interested in them as part of my development. It was an area of record-making that I wanted to learn about, and there was no one better at that point to learn from than Bernard Sumner. Plus, we were both from Manchester and we related to each other, and we had this huge thing in common, which was that we were both looking for respite. It wasn't just about refuge, but it was uncanny that he was looking for respite from the politics of a four-piece Manchester indie group and I was looking for respite from the politics of a four-piece Manchester indie group – it's a simplification, but it was quite uncanny to have that in common. It wasn't all just a reactive thing though. We were both at a point in our lives where we wanted to go for something new, and the touchstones for both me and I think Bernard were David Byrne and Brian Eno – the idea of collaboration and using the studio as an instrument. Back then, that concept had been pioneered by Brian Eno, not only with David Byrne, but also, particularly, with David Bowie on *Low*. So, we thought we were kind of doing that, in our own way. We didn't think about being 'ex-Smiths' and 'ex-New Order' either, but we were constantly being reminded of that wherever we went.

When Bernard and I got together to write, 'rave' hadn't happened yet and the word 'Madchester' was still a few years away. House music was just about to be a thing in everyone's lives. We were just two musicians from bands influenced by Bowie, Eno, Kraftwerk in the studio, but then on the guitar side The Kinks and The Velvet Underground. We both had those things in common. I was in The The at the time, and Bernard was taking a break from New Order. So, I sort of had two day jobs but it worked out fine. During one period, I would get in the car with my roadie on the Monday and drive down to London to work with The The. We'd drive back on the Friday night in time to get to the Haçienda, hang out there and then go into my studio at 3 a.m. to continue writing and recording with Bernard right through until Monday, when it was time to head back to London. That was pretty much what was going on for weeks, and it suited everyone fine. Matt would be working on tracks for *Mind Bomb*, and I'd come down and we were The The, building a serious brotherhood in the band.

With Electronic, Bernard and I didn't know that this tsunami of a new movement was coming along behind us, with lots of other musicians thinking along the same lines – 808 State, the Bristol scene, Mark Moore in S'Express – all these people who were doing this anti-group thing. I was twenty-four at the time, and there was a total revolution going on in my hometown. I'd see Andy Rourke at the Haçienda, other musicians and the friends of musicians – it was just this ridiculously creative explosion. It's an oversimplification, but the things you read and hear about San Francisco in 1966–67, you could draw parallels with the Manchester scene. And not just the obvious drugs and psychedelic thing, but creativity – people getting together during the day and starting up different enterprises in clothes, design and music. Electronic's first album coincided with that.

To Bernard's credit, he made me play more guitar on the record than I had originally intended, which is kind of mad now I think of it. The three big guitar tracks that came out of that period – 'Get The Message', 'Feel Every Beat' and 'Tighten Up' – I'm so glad that they happened. 'Tighten Up' was one that we wrote and recorded sat face to face, Bernard playing my black

Smiths Rickenbacker and me playing Ian Curtis's Vox Phantom. We nailed that song really quickly. I wrote the backing tracks for 'Get The Message' and 'Feel Every Beat' when Bernard was away for a week. Those songs don't sound like either New Order or The Smiths, and I was pleased about that because it was the sound of no one else. It is guitar pop made at a time when pop music didn't have guitars. With 'Get The Message', I was trying to be like The Family Stand's 'Ghetto Heaven'. Before I added the acoustic 12-string, it still worked as a track, like Soul II Soul or something, but adding the guitar really gave it something.

MK Acoustic guitars are an integral part of your sound – something that goes back to the first 'proper' guitar your parents bought you when were eight years old. Both 6- and 12-string acoustics have remained a constant throughout your career. You've always cited Bert Jansch as a major influence. It must've been great to become friends with him and to record the album *Crimson Moon* with him and Bernard Butler.

JM My relationship with Bert Jansch was something that began when I was a kid living on a council estate. He sounded otherworldly, like a wizard, because what he was playing was connected to ancient music. Bert's style was folky and bluesy and also technical. He is one of only a handful of guitarists who's made me actually sit down and think I should really try and work out what he's doing here. I don't usually have the interest or the desire. You see people on YouTube and they can reproduce anything, including my stuff – but I've never been into that. I'm too interested in writing my own songs. But with Bert, I remember listening to some of his albums, like *L.A. Turnaround* and *It Don't Bother Me*, and just thinking, 'How's he done that?' So, becoming his friend was beautiful. Sometimes, you hear something in someone's music and you think that if you were to be friends, you might really get along. And I sort of had that feeling about Bert when I was a kid. There's a sensibility in his music, particularly in some of his instrumental stuff, where you kind of go, 'I really get it, I really get the feeling here.'

He and I got on very well. And studying his playing when I was a kid came in handy when we got to play together, whether it was in his front room, in my kitchen or onstage at the Royal Festival Hall – especially when he went completely off script and started improvising. I just did what I used to do when I was playing along to

his records. It meant a lot to me that he was impressed – 'Yes master, I've put the work in.' We never talked about it, but we felt that playing together should be like a conversation. He was a man of few words, but we used to do all our communicating on the guitar. I'd go round to his place in Kilburn and have a cup of tea, get the biscuits out, and I'd ask him a few questions. He'd ask me questions about what I was doing and then we'd get the two acoustics out and play for about two hours.

I knew Bert's version of 'Angie' before I met my wife, and of all the people that I've played with, all of the freaky, fun things that Angie has seen me do, she has never been so impressed as the day Bert Jansch turned up at our house. Angie is a guitar freak. She hadn't met Bert, but she was a fan of his music. He arrives and, as is his way, like a proper folky, the kettle goes on and we're in the kitchen – always in the kitchen – and we start playing. Angie comes home, and Bert Jansch just happens to be in the kitchen, playing. We stop for a minute and I introduce Bert to Angie and then we carry on playing. His back is to her, and she goes to the kettle and she's just looking at me wide-eyed and mouthing, 'Oh my god, Bert Jansch!' I'm looking up at her going, 'I know!', and Bert is completely unaware. He left a couple of days later, and the first thing Angie said to me was, 'We've been together a long time and that was the best I've ever heard you play.'

MK You joined US band Modest Mouse in 2005. They were the most successful alternative rock band in America at the time. They had a huge audience and were nominated for two Grammy Awards. How was that time? The album you wrote and recorded with them – *We Were Dead Before The Ship Even Sank* – entered the US *Billboard* chart at Number One. Joining that band seems like another pivotal moment.

JM Modest Mouse was another thing that came completely out of the blue. I got a call from the band's leader, Isaac Brock. Their guitarist had left, and he'd been a fan of my playing, so I think Isaac just thought, 'OK, why don't I see if we can get the original guy?' – just on some sort of mad whim, on a hunch. But what's important about the Modest Mouse thing was that a few years earlier I'd become a bit disenchanted with the guitar music I was hearing in England and was listening, almost exclusively, to guitar music from the USA. I met Elliott Smith in Los Angeles and he told me about Portland, Oregon, and mentioned Modest Mouse.

That's where I first heard about them; they intrigued me as a band. So, when I got the call from Isaac, I thought, 'What's all this about?' He just came out and said, 'Do you want to join the band?' And I was like, 'Who is this? I don't even know you!' But there's a line that has run through my life that has always made sense, to me anyway. It could appear quite random to other people, but now that I've been around so long, people can see there's actually a modus operandi. I toId Isaac, 'Well, maybe I'll come over to Portland and hang out with your group for ten days.' That was the arrangement. I was being the same person I was at fourteen or fifteen being asked to play in a band in Manchester. I have this instinct, 'Well, this will be interesting and I've got a feeling that I might be a better guitar player afterwards.' It doesn't matter that I was in The Smiths or The The or whatever, in that scenario I'm the same person I was when I was fifteen. And jumping on a plane and going to Portland is the grown-up version of me getting on a bus to Whalley Range to play with Sister Ray.

The night I arrived, Isaac and I started jamming straight away. My Telecaster wasn't cutting it against his insanely loud Fender Super Six Reverb amplifiers, so I switched to playing his Jaguar and had a life-changing moment. That might sound like an exaggeration, but fifteen years later people are playing Fender Jaguars with my name on them. Within a moment of playing that Jaguar, I'd recognized that something big was going on for me and it wasn't just happenstance. I didn't just fall into it. I woke up the next day and thought to myself, 'While I'm here, I'm going to play that guitar all the time.' In the process of coming up with the riff for 'Dashboard' – which became a really popular song for Modest Mouse and the basis of my relationship with Isaac and the rest of the band – I said to myself, 'This Jaguar is making me play like me.' It was like, 'Where have you been all my life?'

That moment kicked off a new phase in my life and it was all triggered by a riff. Whenever we started playing 'Dashboard', it felt to me like the Jag was the star. It was an exciting time, and on a personal level I don't think it would have happened had I not been in the mindset to have that realization about the guitar. But I was in a place physically, mentally and spiritually where I was open to it.

Some of the Modest Mouse shows we did were some of the best musical moments I've had. We'd start improvising and then go off, and it was totally unique and also pretty heavy. Modest Mouse live was heavy music. It would sometimes derail into wildness and chaos, which was all part of it. We had a Number One album

in the USA, which felt like a victory for alternative music. I thought of it as a victory for the guitar.

MK Your discovery of the Jaguar in 2005, while playing with Modest Mouse, led to you becoming the guitarist most associated with that particular Fender guitar. Fender then invited you to design your own Johnny Marr signature Jaguar model, which has remained in production since 2012. Around the same time, you joined The Cribs and toured extensively with them, as well as co-writing and recording the 2009 album *Ignore The Ignorant*. Your signature Jaguar has been a really successful collaboration for both you and Fender. It must've been a labour of love.

JM Designing my signature Jag was a total obsession; it had to be. I took that with me into The Cribs, which they'll remember well. Being in The Cribs was great. It happened because we got together to write a few songs and the creative process just took off. We wrote a really good set of songs and the shows were explosive. I was doing a lot of tinkering with my prototype Jag at the time – backstage and sometimes even onstage. There was a gig in Toronto, near the end of The Cribs' North American tour, where I had a screwdriver on my amp and I was changing the height of the bridge saddles in between songs. I was convinced that the bridge was dropping, and it being a prototype it might have been. But I don't think it could have dropped during the length of a song, especially as The Cribs songs are quite short. I became obsessive about getting the guitar right, but it paid off in the end. I see it as an honour having your name on a guitar, like Chet Atkins or Les Paul. I know that there are lots of signature models out there these days, but I took mine very seriously.

There was a phrase that kept coming into my mind: 'unwanted conditions' – technical things like the bridge dropping and a switch that could turn off the guitar if you hit it by accident. In 2010, you had no need to switch off the guitar. The reason some guitars were designed with an off switch was primarily because of radio frequency interference on radio dates with orchestras in places like Chicago and New York, where there were train lines near the studios. The guitar player might not actually be playing until bar 124, but the guitar might be getting all the interference noise, so he'd use the kill switch. By 2010, there was no need for that.

I have to give a lot of credit to my friend Bill Puplett, who is an amazing luthier in London and who has been working with me since 1988. He really helped me: building prototypes, listening to my ideas and giving me advice. Bill suggested the Bare-Knuckle pickups, and we devised all sorts of innovations, including a tremolo arm that stays where you leave it. We changed almost everything there is to change on the Jaguar, except the headstock and machine heads. The sound ended up as a culmination of almost everything I do on records. We made it so it rings out like a Rickenbacker or a clean Gibson Les Paul, and chimes like a Gretsch. Essentially, it sounds like three of my old guitars all at once. I'm able to do it all with my one guitar.

When it came to the colour of the Jag, the first ones were either Olympic White or Metallic KO, which is an original colour that was copied from an old faded-out Candy Apple Red that had oxidized and looked great. Then I did a Sherwood Green, which was really popular. I'm not adding a new colour for commercial purposes, or to make them more collectable; it's just part of what happens. When I feel like it's time to have a new colour, then I'll do it. With the most recent, Fever Dreams Yellow, I'd been touring with the Comet Sparkle for quite a long time and that guitar became synonymous with the *Call the Comet* tour, plus I'd played it on the theme song for the Bond film, so I thought, 'It's been three and a half years. It's done loads of touring and then the Bond theme, so it's time for a change.'

The *Fever Dreams* record and tour felt like a different period in my life, so when I'm playing and I'm looking down and having this relationship with my instrument, I don't want it to put my head back a few years. I think all guitarists are a bit like this; every few years, you want to mix it up. The difference is, though, that my signature Jag is perfection for me as an instrument. So, right now, I'm simply changing the colour.

These days, I quite like the fact I don't really change guitars much onstage. There are a couple of songs where I've got the sustainers going, because that sound has become a thing in my songs over the past few years.

In the studio, I still mix it up a lot and pull out different guitars that make me do different things. The Les Paul Customs have been a big part of my last two solo albums, and the double-neck 12-string has been on the movie soundtracks. If I'm going to guest onstage with someone like Primal Scream or Alicia Keys, I'm always going on with my signature Jag. That says it all really.

Over the years, I've found a way of working with effects, especially on the movie soundtracks with Hans Zimmer. Being able to manipulate the effects with my right hand and play the sustainer with my left – I'm not sure it's something sound designers can do with a synthesizer. There are things a guitar can do that keyboards can't, because you're dealing with strings, metal, magnets and feedback. You can get much more physical and really get a human response from a guitar. You can be clever and try to design that with a synth, but with a guitar you're actually physically pulling wires. So, that's an opportunity that my career has presented that I couldn't have imagined. I'm not sure the twenty-three-year-old me could have done what I do with Hans – standing in front of an orchestra and a thirty-piece choir. I just wouldn't have had the mindset.

MK You finally released your first solo album *The Messenger* in 2013 to critical acclaim and commercial success. Now, after four solo albums, it's hard to imagine a time when Johnny Marr wasn't the frontman of his own band.

JM In many ways, it feels like all the various bands and projects I've been involved with over the years were the steps I needed to take – varied and important ones – to get where I am now. Reviews of my solo records sometimes say they can hear bits of Electronic, The Smiths, The The or whoever, and that's fine. It's a compliment and makes sense. I haven't dropped the acoustic techniques I was using in The Smiths. If you listen to something like 'European Me' or 'New Town Velocity' off *The Messenger* album, you hear all that. I guess another side of that is that the title track from *Playland* is not unlike what I was doing in The The. Whether it's 'New Town Velocity' or 'Easy Money', which reminds people of what I was doing in Modest Mouse, it all makes sense to me. It's all what I do, and I feel like it's all been leading up to where I am now.

It might sound obvious, but it almost feels like a duty now for me and my band to always be known as a guitar band. I'm always going to try and be that, and it goes back to how this whole thing started – with me hearing Marc Bolan, Mick Ronson, James Williamson, Rory, Bert Jansch, all of it really.

As far back as 1983, I was being asked the question, 'Is guitar music over?' In those early days, it was, 'Are you railing against synth and electronic music?' And then again in 1987–88, it was all, 'Do you think dance music has killed off the guitar?' I still get asked that question; I probably always will. I just say, 'You're talking to the wrong person.'

YAMAHA SG-1000
Brown Sunburst
1983
#097514

YAMAHA SG-1000
Deep Blue
1983
#110972

YAMAHA SG-1000
Cherry Sunburst
1981
#057863

YAMAHA SG-700
Brown
1977
#001573

YAMAHA SG-1300
Burgundy
1983
#122940

When I was kicking around in my teenage bands, Yamahas were the guitar of choice with some of the more interesting guitar players of the time – people like Bill Nelson and John McGeoch. They became popular because, stylistically, Yamahas meant a break away from the classic rock tradition of the past and also because, as an instrument, they are really very good. ∎

235

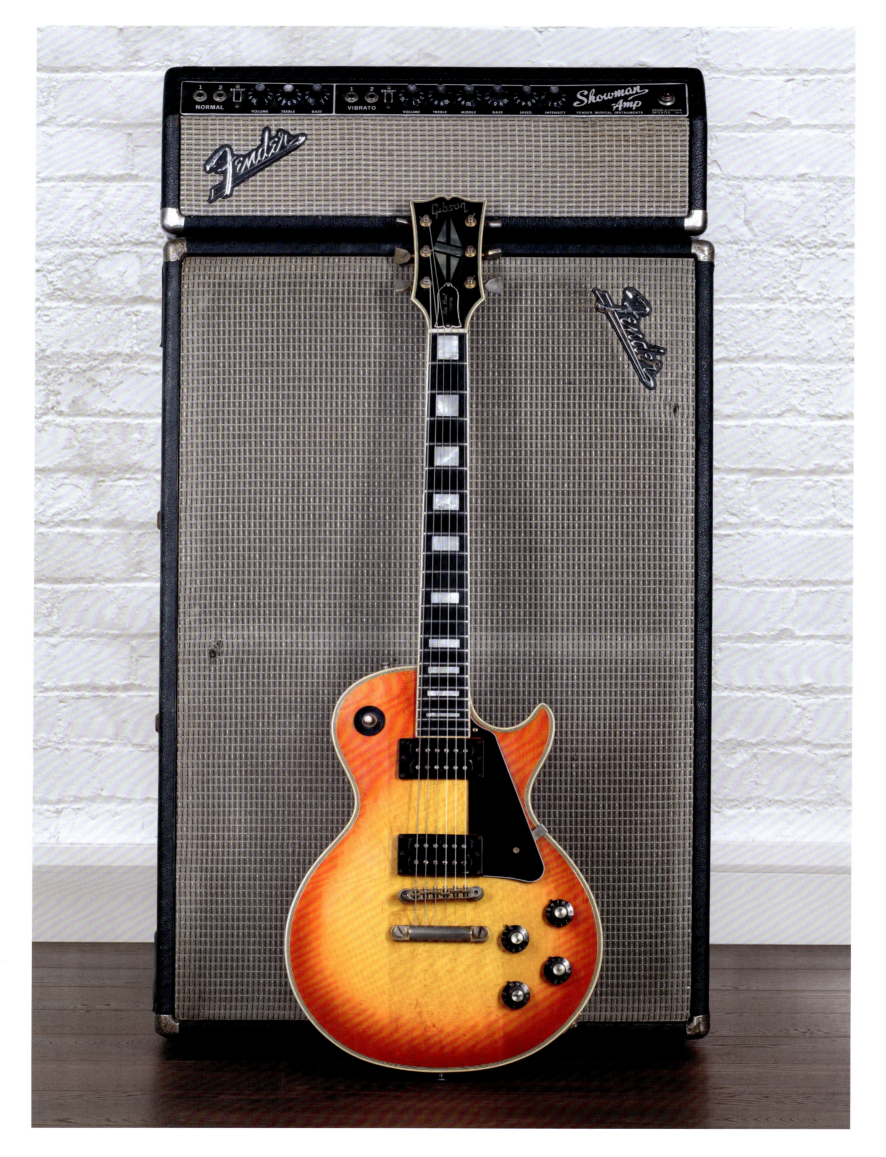

GIBSON LES PAUL CUSTOM
Sunburst
1973
#204871

One of my favourite guitar players is James Williamson. The first time I heard him was on the album *Raw Power* by Iggy and the Stooges. His riffs, attitude and sound were amazing. I'm not known for sounding like him, although I did in a couple of my early bands. I do incorporate things I've learnt from everywhere, and these guitars are perfect for delivering a heavy attitude. ∎

GIBSON LES PAUL CUSTOM
Black
1973
#125272

This is the backup and alternative for the sunburst Les Paul Custom (on the previous page). I've used both a lot on my solo albums. I like these 1973 Les Paul Customs because they're the sound of British punk rock in the 1970s. ∎

**FENDER
PRECISION BASS**
Lake Placid Blue
1965
#101188

FENDER
PRECISION BASS
Sunburst
1965
#100345

'DASHBOARD' DECK GUITAR
Natural Wood
c. 2007
NO SERIAL NUMBER

This was built for me for the Modest Mouse 'Dashboard' video in 2007

FENDER TELECASTER
Butterscotch Blonde
2017
#R16707

Given to me by Chrissie Hynde in 2019. Thank you, Chrissie. ∎

FENDER STRATOCASTER – NILE RODGERS HITMAKER STRATOCASTER
Olympic White Custom Shop
2014
#NR0004

While I was on tour with The Cribs and still in the process of building my Signature Jaguar prototype, I had the idea that I'd like to give one of the first ones to Nile Rodgers, as a thank you for being an inspiration. In 2012, I got the opportunity to present Nile with one of my guitars. It was a nice moment. Sometime later, when I was about to go onstage in New York, Nile appeared in my dressing room with this Strat, which he then gave to me in return. ∎

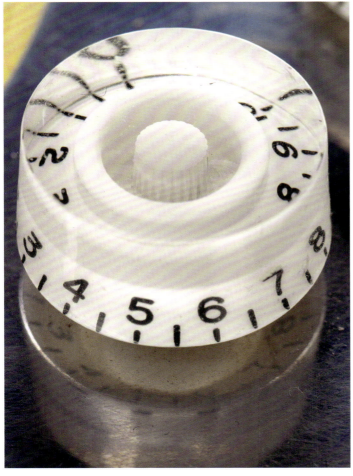

SILVERTONE 1415
Bronze
1961
NO SERIAL NUMBER

One of my biggest heroes as a kid was the late great Rory Gallagher, who I saw in concert many times. He was a crucial figure in my development as a musician, not only for his music but for the example he set. He made me think that all I would need in the world was a guitar and an amplifier. He was good to his fans, and I got to know him very briefly before his untimely death in 1995. A couple of years later, his management and family contacted me to say they wanted to give me this guitar, which he'd used on 'Cradle Rock' and 'A Million Miles Away'. ∎

GIBSON ES-175
CUSTOM SHOP
STEVE HOWE MODEL
Sunburst
2002
#00472719

GIBSON ES-175
Sunburst
1963
NO SERIAL NUMBER

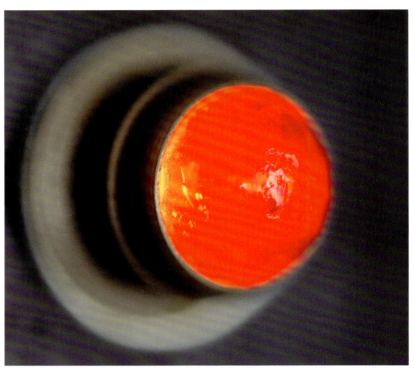

HAGSTROM EDP46 DE LUXE
Blue Sparkle
c. 1963
NO SERIAL NUMBER

One of the longest-running collaborations throughout my career has been with Bryan Ferry, who I first worked with in 1987 and who I continue to work with to this day. One of the many iconic images of Bryan is from the gatefold sleeve of the 1973 Roxy Music album *For Your Pleasure*, where he is holding this Hagstrom guitar. Very few people knew what a Hagstrom guitar was before this picture (below) made it famous and, as you can see, its colour has faded with time. Bryan gave me the guitar in 2018 as a gift. It's one of my most treasured items. ∎

GIBSON EDS-1275 DOUBLE NECKS
Cherry / Black
1993 / 1994
#93033326 / #19667

I first started working with Hans Zimmer in 2010 on the soundtrack for the movie *Inception*, and since then we've worked on a few more together. I wanted a very specific 12-string sound for *Inception*, and when I came across the Double Neck it became my signature sound for the film. As well as its distinctive look, the guitar has a distinctive sound due to its large body mass and darker-, more 'rock'-sounding pickups, which makes for a great electric 12-string sound. Working on movies, I have to come up with a lot of different ideas, so having both 6 strings and 12 strings in my hands is almost like playing two guitars at once. It also means I can play in two different tunings without having to change guitars. ■

**FENDER JAGUAR –
JOHNNY MARR SIGNATURE**
Fever Dreams Yellow
2021
#V2095956

This is the most recent of my custom colour signature Jaguars. It was made for the *Fever Dreams Pts 1–4* album and then played on tour with Blondie and The Killers. ∎

RICKENBACKER 340
Jetglo
1984
#XD0404

In 2022, I was invited to go on a tour of the USA and Canada with The Killers. I had the idea to try to find a Rickenbacker because The Killers had asked me to play some Smiths songs with them in their set. I went into rehearsals and didn't give it any more thought. A few days before the tour started, a friend of mine called to say that there was someone selling a Rickenbacker that I had supposedly signed sometime in the 1990s. I actually guessed which guitar it was, because in 1995 Bernard Sumner and I were on a promotional tour with Electronic and a guy had asked me to sign his Rickenbacker. I remembered it as it was the first guitar I'd ever signed. Back then, it wasn't so common for people to sign guitars. We contacted the person, who was in Canada, and told him that I would buy it. I was starting the tour in Vancouver, so he was able to bring it to the first show. When I got the guitar, it still had my signature on it, which I removed before using it with The Killers in their set. ■

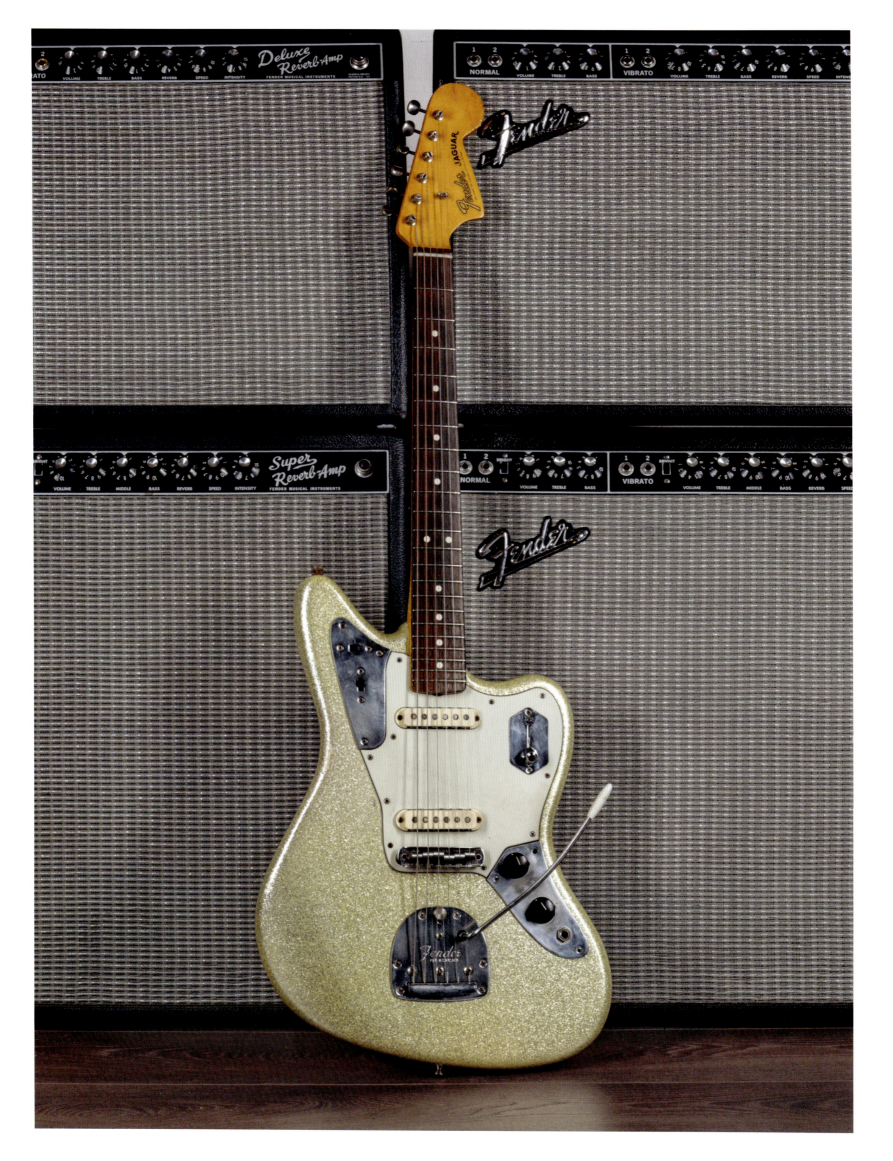

FENDER JAGUAR –
JOHNNY MARR SIGNATURE
Comet Sparkle
2018
#V210532

At the start of the *Call The Comet* album, I asked Fender if they could make me a Signature Jaguar with what's usually called a 'Champagne Sparkle' finish. They experimented in the Custom Shop, and the result was this one-of-a-kind guitar, which we called 'Comet Sparkle' and which I used for my shows during 2017 and 2018. When Hans Zimmer asked me to play guitar on the soundtrack to the James Bond movie *No Time To Die* in 2018, I used this guitar to play the 'James Bond Theme'. I then used it on the title song, 'No Time To Die' with Billie Eilish, which won the award for Best Original Song at the Oscars in 2022. ■

IMAGE CAPTIONS

— ENDPAPERS: The The Versus The World tour, 1990

002 *The Queen Is Dead* tour, North America, 1986. Photograph by Nalinee Darmrong

008 Soundcheck, Inverness, 1 October 1980. Photograph by Angie Marr

011 Wythenshawe, Manchester, 1981. Photograph by Angie Marr

013 Electric Ballroom, Camden, London, 1983. Photograph by Paul Slattery, Camera Press London

015 Gretsch Super Axe, British Pop Archive. John Rylands Research Institute and Library, Manchester, 2023. Courtesy The University of Manchester

021 GLC Jobs For A Change Festival, Jubilee Gardens, London, 1984. Photograph by Paul Rider, Camera Press London

026 *NME* cover shoot, London, 2013. Photograph by David Edwards

029 Stills from 'Heaven Knows I'm Miserable Now' performance, *Top of the Pops*, London, 1984. BBC Motion Gallery / Getty Images

032 Stills from 'Nowhere Fast' performance, *The Old Grey Whistle Test*, London, 1984. BBC Motion Gallery / Getty Images

035 Nevern Square, Earl's Court, London, 1984. Photograph by Angie Marr

047 Stills from 'The Boy With The Thorn In His Side' performance, *Top of the Pops*, London, 1985. BBC Motion Gallery / Getty Images

050 L'Eldorado, Paris, 1984. Photograph by Paul Slattery, Camera Press London

054–55 Stills from 'Dogs Of Lust' video, London, 1993. Photographs by Lawrence Watson

061 Writing 'Nowhere Fast', Nevern Square, Earl's Court, London, 1984. Photograph by Angie Marr

065 Blackpool, 20 June 1984. Photograph by Angie Marr

067 *Record Mirror* (14 June 1986), London. Photograph by Pat Bellis

071 Oxford, 18 March 1985. Photograph by Andrew Caitlin

081 Noel Gallagher, 100 Club, London, 1994. Photograph by Paul Slattery, Camera Press London

083 *The Queen Is Dead* tour, North America, 1986. Photograph by Nalinee Darmrong

087 Noel Gallagher, The Astoria, London, 1994. Photograph by Peter Macdiarmid / Getty Images

088 *The Queen Is Dead* tour, North America, 1986. Photograph by Nalinee Darmrong

091 *The Queen Is Dead* tour, North America, 1986. Photograph by Nalinee Darmrong

096 Stills from 'Shoplifters Of The World Unite' performance, *The Tube*, Newcastle, 1987. © ITV Studios Ltd; images reproduced with the permission of ITV Studios

101 Recording 'Death At One's Elbow', The Wool Hall, Beckington, Somerset, 1987. Photograph by Angie Marr

103 The Coburg Hotel, London, 1984. Photograph by Daniel Lainé

115 *Playland* tour, The Forum, London, 2015. Photograph by Pat Graham

117 *Guitar Player* (January 1990), Roland Gardens, London. Photograph by Todd Fath / *Guitar Player* / Future US LLC

119 Bowdon, Cheshire, summer 1987. Photograph by Angie Marr

123 Stills from 'Hi Hello' video, Rochdale and Manchester, 2018. Directed by Mat Bancroft and Johnny Marr; courtesy Sitcom Soldiers

127 London, Ontario, 1986. Photograph by Angie Marr

137 Onstage with Electronic, Dodger Stadium, Los Angeles, 1990. Photograph by Kevin Cummins / Getty Images

157 Stills from 'Spirit Power And Soul' video, Manchester, 2021. Directed by Mat Bancroft and Johnny Marr; courtesy Sitcom Soldiers

161 *Guitar Player* (February 2003), Soho, New York. Photograph by Stephanie Pfriender Stylander / Guitar Player / Future US LLC

169 Bert Jansch with Johnny Marr, Kilburn, London, 2000. Photograph by Loren Jansch, used by permission of Loren Jansch's estate

170 Onstage with The Healers, Melkweg, Amsterdam, 31 March 2003. Photograph by Peter Pakvis / Redferns

175 Onstage with R.E.M., Madison Square Garden, New York, 2008. Photograph by Chris Sikich

178 'Happy Birthday Fender' 60th anniversary Fender advert, Urbis, Manchester, 2006. Fender® / Mick Rock

181 Backstage, Modest Mouse, 2007. Photograph by Pat Graham

191 Onstage with Modest Mouse, 2008. Photograph by Pat Graham

193 Onstage with Modest Mouse, 2008. Photograph by Pat Graham

199 Coming offstage, Philadelphia, 2008. Photograph by Pat Graham

202 'This is the sound' Fender advert, The Ritz, Manchester, 2013. Fender® / Shirlaine Forest

215 *Playland* tour, Brixton Academy, London, 2014. Photograph by Pat Graham

220–21 'Fender Johnny Marr Jaguar Signature Model information schematic', 2012. Created by Paul Kelly

243 Still from 'Dashboard' video, Los Angeles, 2007. Video directed by Grady Hall; turntable guitar designed by production designer Teri Whittaker. Courtesy of Epic Records, a division of Sony Music Entertainment

247 Onstage with Nile Rodgers & Chic, The Ritz, Manchester, 2013. Photograph by Elspeth Mary Moore

251 Rory Gallagher recording 'Cradle Rock' and 'A Million Miles Away', London, 1973. Photograph by Michael Putland © Michael Putland Archive

265 Roxy Music *For Your Pleasure* inner-gatefold, 1973. Courtesy of EMI, a division of Universal Music Operations Ltd

270 Recording the *No Time To Die* soundtrack, Soho, London, 2020. Photograph by Angie Marr

272 The Crazy Face Factory, Manchester, 2022. Fender® / Dan Massie

277 Onstage with The Killers, North American tour, 2022. Photograph by Rob Loud

279 Onstage with Billie Eilish, The Brit Awards, London, 2020. Photograph by Samir Hussein / WireImage / Getty Images

280 Recording the 'James Bond Theme' for the *No Time To Die* soundtrack, 2020. In the photos with Johnny Marr at AIR Studios, London, are: (top) Paul Willey (violin), Natalia Bonner (violin); (bottom) Matt Dunkley (conductor), Steve Morris (violin), Natalia Bonner (violin), Warren Zielinski (violin), David Lale (cello). *NO TIME TO DIE* © 2021 Danjaq, LLC and Metro-Goldwyn-Mayer Studios Inc. All Rights Reserved.

288 Soho, New York, 2003. Photograph by Stephanie Pfriender Stylander

— All guitar portraits and macro photographs by Pat Graham, taken at The Crazy Face Factory, 2022.

GUITARS BEHIND THE TRACKS

GRETSCH SUPER AXE (012)
Used to record: The Smiths 'Hand In Glove'
Used to write: The Smiths 'Hand In Glove', 'What Difference Does It Make?', 'These Things Take Time', 'Reel Around The Fountain', 'Jeane', 'This Charming Man'

RICKENBACKER 330 (016)
Used to record: The Smiths 'What Difference Does It Make?', 'This Charming Man', 'Still Ill', 'Accept Yourself', 'Pretty Girls Make Graves', *The Smiths*; Electronic *Raise The Pressure*; The Cribs *Ignore The Ignorant*; Johnny Marr 'This Tension'

GIBSON ES-355 (024)
Used to record: The Smiths 'Heaven Knows I'm Miserable Now', 'Barbarism Begins At Home'; Sandie Shaw 'Hand In Glove'; The Pretenders 'Windows Of The World'

FENDER TELECASTER GIFFIN CUSTOM KORINA (028)
Used to record: The Smiths 'Girl Afraid', 'Meat Is Murder', 'Nowhere Fast', 'The Headmaster Ritual'

GIBSON J-160E (034)
Used to record: The Smiths 'Please, Please, Please Let Me Get What I Want', 'William, It Was Really Nothing'

EPIPHONE CORONET (038)
Used to record: The Smiths 'William, It Was Really Nothing', 'The Headmaster Ritual', 'You Just Haven't Earned It Yet, Baby'; Johnny Marr 'The Messenger'

OVATION LEGEND 1867 (040)
Used to perform: The Smiths 'Please, Please, Please Let Me Get What I Want', 'William, It Was Really Nothing', 'Rusholme Ruffians'

GIBSON LES PAUL STANDARD (046)
Used to record: The Smiths 'The Headmaster Ritual', 'That Joke Isn't Funny Anymore','What She Said'; The Pretenders '1969'; The The 'Slow Emotion Replay'; New Order 'Regret'; The Cribs 'We Share The Same Skies'; Johnny Marr 'Generate! Generate!', 'Bug'; Noel Gallagher's High Flying Birds 'Pretty Boy'

MARTIN D-28 (056)
Used to record: The Smiths 'Well I Wonder', 'Cemetry Gates', 'Half A Person', 'There Is A Light That Never Goes Out'; Johnny Marr 'European Me'

EPIPHONE CASINO (060)
Used to record: The Smiths 'Nowhere Fast', 'How Soon Is Now?'

FENDER STRATOCASTER (LEFT) (066)
Used to record: The Smiths 'Well I Wonder', 'The Boy With The Thorn In His Side', 'Shakespeare's Sister', 'Shoplifters Of The World Unite'; Bryan Ferry 'The Right Stuff'; Talking Heads 'Ruby Dear'; Stex 'Still Feel The Rain'

FENDER STRATOCASTER (RIGHT) (066)
Used to record: The Smiths 'There Is A Light That Never Goes Out', 'Unlovable', 'Rubber Ring', 'Half A Person', 'How Soon Is Now?', 'Girlfriend In A Coma', 'Unhappy Birthday'

GRETSCH 6120 SINGLE CUT (076)
Used to record: The Smiths 'Ask', 'Is It Really So Strange?', 'Stretch Out And Wait'; *The Amazing Spider-Man 2 (The Original Motion Picture Soundtrack)*
Used in the video for: The Smiths 'Panic'

GIBSON LES PAUL STANDARD (078)
Used to record: The Smiths 'Panic', 'London', 'Sweet And Tender Hooligan'

GIBSON LES PAUL CUSTOM (084)
Used to record: The Smiths 'The Queen Is Dead', 'Bigmouth Strikes Again', 'I Started Something I Couldn't Finish', 'Last Night I Dreamt That Somebody Loved Me'; Electronic 'Feel Every Beat'; Oasis '(Probably) All In The Mind'

FENDER STRATOCASTER (090)
Used to record: The Smiths 'I Know It's Over', 'Some Girls Are Bigger Than Others'

GRETSCH COUNTRY CLUB (094)
Used to record: The Smiths 'Shoplifters Of The World Unite'

GIBSON ES-335/12 (096)
Used to record: The Smiths 'Half A Person', 'Sheila Take A Bow', 'Paint A Vulgar Picture', 'Stop Me If You Think You've Heard This One Before'; Talking Heads '(Nothing But) Flowers', 'Cool Water', 'Ruby Dear'; Electronic 'Get The Message'

FENDER BASS VI (100)
Used to record: The Smiths 'Death At One's Elbow'; The The 'Helpline Operator'; The Cribs 'Save Your Secrets', 'Stick To Yr Guns'

MARTIN D-41 (102)
Used to record: The Smiths 'Unhappy Birthday', 'Stop Me If You Think You've Heard This One Before', 'Girlfriend In A Coma'; Billy Bragg 'Sexuality'; The The 'Love Is Stronger Than Death' (played by Matt Johnson); Pet Shop Boys 'Did You See Me Coming?'

MARTIN D-28/12 (110)
Used to record: The Smiths 'Half A Person', 'Unhappy Birthday'; Electronic 'Get The Message', 'Can't Find My Way Home'; Lisa Germano 'Paper Doll'

FENDER STRATOCASTER (116)
Used to record: The Smiths 'Paint A Vulgar Picture'

FENDER TELECASTER (118)
Used to record: Billy Bragg 'Sexuality', 'Cindy Of A Thousand Lives'; Electronic 'Forbidden City'

GIBSON ES-335/12 (CENTRE) (122)
Used to record: The Pretenders 'Windows Of The World'

GIBSON ES-335/12 (RIGHT) (122)
Used to record: Johnny Marr 'Hi Hello'

FENDER STRATOCASTER (136)
Used to record: The The 'The Beat(en) Generation', 'Jealous Of Youth'; Kirsty MacColl 'Days', 'Children Of The Revolution', 'Walking Down Madison'; Pet Shop Boys 'This Must Be The Place I Waited Years To Leave', 'My October Symphony'; Electronic 'Getting Away With It', 'Idiot Country', 'Disappointed'

GIBSON LES PAUL CUSTOM (140)
Used to record: Electronic 'Make It Happen', 'When She's Gone', 'Haze'; The Healers 'InBetweens'

GIBSON LES PAUL SPECIAL (142)
Used to record: Electronic 'Second Nature'; The Cribs 'Cheat On Me', 'Ignore The Ignorant'

GIBSON LES PAUL GOLD TOP (146)
Used to record: Radiohead *In Rainbows*; Johnny Marr *Playland*, *Call The Comet*

GIBSON S-1 (153)
Used to record: Johnny Marr *The Messenger*, *Playland*

GIBSON SG STANDARD (LEFT) (160)
Used to record: Electronic 'Vivid'; Oasis 'Born On A Different Cloud'; The Healers 'The Last Ride', 'Caught Up', 'You Are The Magic', 'Long Gone'

GIBSON SG STANDARD (RIGHT) (160)
Used to record: Pet Shop Boys 'I Get Along', 'Love Is A Catastrophe', 'I Didn't Get Where I Am Today'; Radiohead *In Rainbows*
Used to perform: with Neil Finn in 7 Worlds Collide

YAMAHA LLX-400 (166)
Used to record: Electronic 'Second Nature'; Bert Jansch 'Fool's Mate', 'The River Bank'; Jane Birkin 'Mother Stands For Comfort', 'Waterloo Station'; Pet Shop Boys 'Did You See Me Coming?', 'Beautiful People'
Used to perform: with Bert Jansch

RICKENBACKER 370/12 (170)
Used to record: The Healers 'Down On The Corner'; Tweaker 'The House I Grew Up In'; Modest Mouse 'Missed The Boat'

GRETSCH 6120 DOUBLE-CUTAWAY (176)
Used to perform: with Neil Finn in 7 Worlds Collide

**FENDER TELECASTER CUSTOM
SHOP RELIC** (178)
Used to record: Lisa Germano 'Into The Night'; Johnny Marr 'Easy Money'

FENDER JAZZMASTER (182)
Used to record: Modest Mouse 'Steam Engenius', 'Fly Trapped In A Jar', 'Education'

FENDER JAZZMASTER CUSTOM (184)
Used to record: Modest Mouse 'Fire It Up', 'Missed The Boat', 'People As Places As People'

FENDER JAZZMASTER (187)
Used to record: The Cribs *Ignore The Ignorant*

FENDER JAZZMASTER (190)
Used to perform: Modest Mouse 'Tiny Cities Made Of Ashes', 'Bury Me With It', 'Spitting Venom'

FENDER JAGUAR (192)
Used to record: Modest Mouse 'Dashboard'; Pet Shop Boys 'Did You See Me Coming?', 'Pandemonium'; Johnny Marr 'Upstarts'

FENDER JAGUAR (198)
Used to record: Modest Mouse 'We've Got Everything', 'Invisible', 'Florida'

FENDER JAGUARS (202)
Used to record: The Cribs 'Cheat On Me', 'Last Year's Snow'; *Inception (Music From The Motion Picture)*; Johnny Marr 'The Right Thing Right', 'European Me', 'Lockdown', 'New Town Velocity'

FENDER JAGUAR – JOHNNY MARR SIGNATURE (209)
Used to record: Johnny Marr 'Easy Money', 'Candidate', 'I Feel You'; Bryan Ferry 'Soldier of Fortune'; Chris Spedding 'Heisenberg'; Noel Gallagher's High Flying Birds 'Ballad Of The Mighty I'

FENDER JAGUAR – JOHNNY MARR SIGNATURE (216)
Used to record: Johnny Marr 'Walk Into The Sea', 'Hi Hello'; Johnny Marr and Maxine Peake 'The Priest', Noel Gallagher's High Flying Birds 'If Love Is The Law', 'Pretty Boy'; The Avalanches 'The Divine Chord'; *No Time To Die (Original Motion Picture Soundtrack)*

FENDER STRATOCASTER (222)
Used to record: The Healers 'Caught Up'

YAMAHAS (234)
Used to record: Johnny Marr 'Dynamo', '25 Hours', 'Jeopardy'; The The 'We Can't Stop What's Coming'

GIBSON LES PAUL CUSTOM (235)
Used to record: Johnny Marr 'Hey Angel', 'Rise', 'Tenement Time'

FENDER PRECISION BASS (240)
Used to record: Johnny Marr 'The Speed Of Love', 'Lightning People', 'Night And Day'

FENDER PRECISION BASS (241)
Used to record: Johnny Marr 'Human'

GIBSON EDS-1275 DOUBLE NECKS (270)
Used to record: *Inception (Music From The Motion Picture)*; *The Amazing Spider-Man 2 (The Original Motion Picture Soundtrack)*; *Freeheld (Original Motion Picture Soundtrack)*; *No Time To Die (Original Motion Picture Soundtrack)*

FENDER JAGUAR – JOHNNY MARR SIGNATURE (272)
Used to record: Johnny Marr *Fever Dreams Pts 1–4*

FENDER JAGUAR – JOHNNY MARR SIGNATURE (278)
Used to record: Johnny Marr 'The Tracers', 'Spiral Cities', 'A Different Gun'; Billie Eilish 'No Time To Die', 'James Bond Theme'

INDEX

ACKNOWLEDGMENTS

MARR'S GUITARS
by Johnny Marr

PHOTOGRAPHY
Pat Graham

ART DIRECTION
Mat Bancroft

Thank you

Tristan de Lancey, Jane Laing, Nick Jakins, Jane Cutter
and Florence Allard at Thames & Hudson

Brian Message, Sammi Wild at ATC Management

Elizabeth Sheinkman at Peters Fraser + Dunlop

Andy Booth at SAS Daniels

Angie Marr, John and Frances Maher, Bill Puplett, Mat Bancroft,
Pat Graham, Richard Henry, Martin Kelly, Nile Marr, Sonny Marr,
Sophie Putland, Alan Rogan, Justin Norvell, David Mulqueen,
Neil Whitcher, Joe Moss, Dave Cronen, Matt Johnson, Hans Zimmer,
Bryan Ferry, Bernard Butler, Noel Gallagher, Ed O'Brien, Andy Rourke,
Isaac Brock, Zak Starkey, Bernard Sumner, Neil Tennant, Chris
Lowe, Neil Finn, J Mascis, Billy Duffy, Ryan Jarman, Gary Jarman,
Ross Jarman, Charlie Burchill, Barrie Cadogan, Peter Buck,

The Edge, Chrissie Hynde, Karen Kitson, Michael Eastwood,
Brian Eastwood, Rick Zsigmond, Jamie Franklin, Ken Nelson,
Phil Manzanera, Daniel Gallagher, Donal Gallagher, James
Doviak, Jack Mitchell, Iwan Gronow, Clive Brown, Spenny,
Rich House

To the roadies and techs who've been by my side in the trenches
and on the front lines over the years and to all the guitar gods
who've inspired me throughout my life, thank you:

Marc Bolan, James Honeyman-Scott, Rory Gallagher,
James Williamson, Bert Jansch, Nils Lofgren, Joni Mitchell,
Nile Rodgers, Mick Ronson, Chris Spedding, John McLaughlin,
George Harrison, Pete Townshend, Bill Nelson, Keith Richards,
Brian Jones, Richard Lloyd, Scotty Moore, Sterling Morrison,
John McGeogh, Bo Diddley, Chet Atkins, John Perry, Jerome
Smith, Tony Hicks and Johnny Thunders.

Front cover: Johnny Marr backstage at
The Smiths gig at the University of East Anglia,
Norwich, 14 February 1984. Photograph by
Paul Slattery, Camera Press London

Back cover: Three details of Rickenbacker 330,
Jetglo, 1982, #VI2485. Photographs by Pat Graham

First published in the United Kingdom in 2023
by Thames & Hudson Ltd, 181A High Holborn,
London WC1V 7QX

Reprinted 2024

Marr's Guitars © 2023 Thames & Hudson Ltd,
London

Text by Johnny Marr © 2023 Marr Songs Ltd

Foreword © 2023 Hans Zimmer
Studio photographs by Pat Graham © 2023
Marr Songs Ltd

For image copyright information, see page 284

British Library Cataloguing-in-Publication Data
A catalogue record for this book is available from
the British Library

ISBN 978-0-500-02632-8

Printed and bound in China by C&C Offset
Printing Co. Ltd

Be the first to know about our new releases,
exclusive content and author events by visiting
thamesandhudson.com
thamesandhudsonusa.com
thamesandhudson.com.au

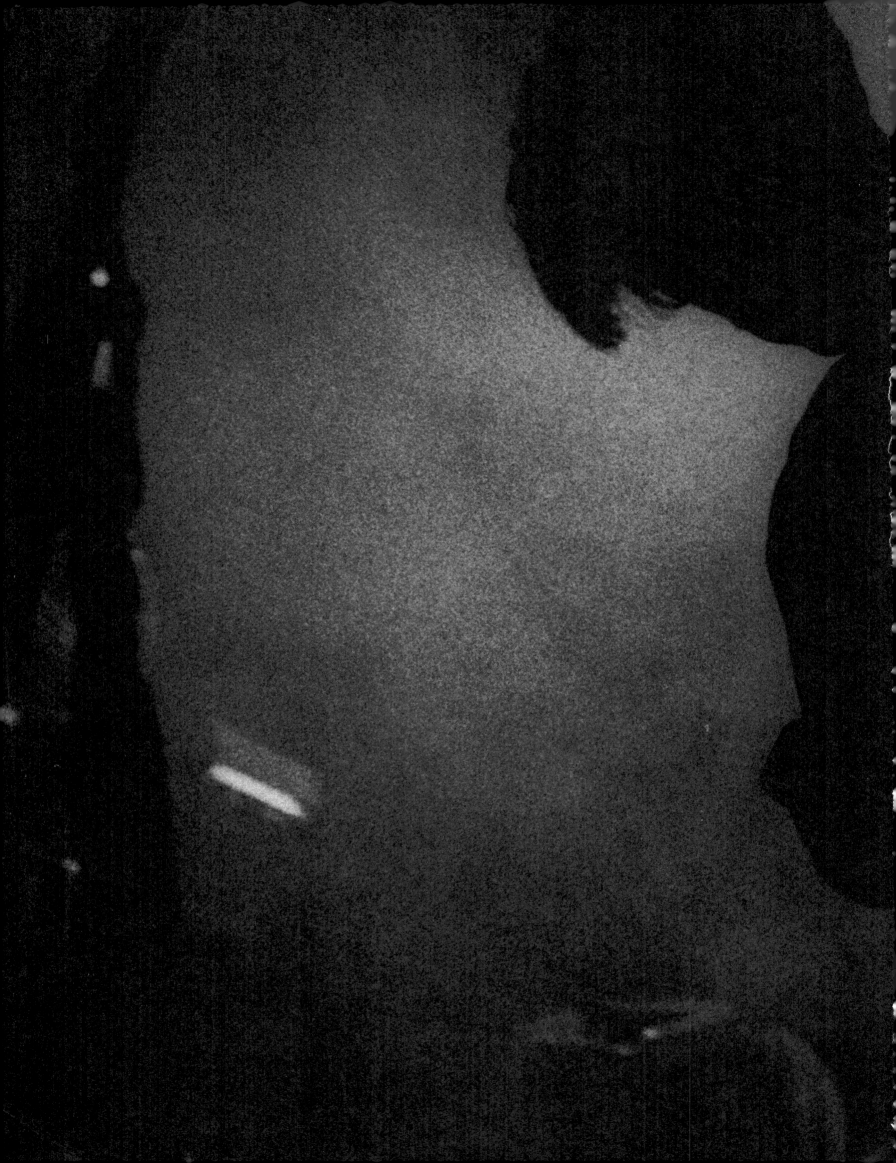